INSIDE THE CITADEL

Also by Richard Symonds

THE MAKING OF PAKISTAN

THE BRITISH AND THEIR SUCCESSORS

INTERNATIONAL TARGETS FOR DEVELOPMENT (*editor*)

THE UNITED NATIONS AND THE POPULATION QUESTION (*with Michael Carder*)

OXFORD AND EMPIRE

ALTERNATIVE SAINTS: The Post-Reformation British People Commemorated by the Church of England

FAR ABOVE RUBIES: The Women Uncommemorated by the Church of England

Inside the Citadel

Men and the Emancipation of Women, 1850–1920

Richard Symonds

First published in Great Britain 1999 by
MACMILLAN PRESS LTD
Houndmills, Basingstoke, Hampshire RG21 6XS and London
Companies and representatives throughout the world

A catalogue record for this book is available from the British Library.

ISBN 0–333–73733–4

First published in the United States of America 1999 by
ST. MARTIN'S PRESS, INC.,
Scholarly and Reference Division,
175 Fifth Avenue, New York, N.Y. 10010

ISBN 0–312–22202–5

Library of Congress Cataloging-in-Publication Data
Symonds, Richard, 1918–
Inside the citadel : men and the emancipation of women, 1850–1920
/ Richard Symonds.
p. cm.
Includes bibliographical references and index.
ISBN 0–312–22202–5 (cloth)
1. Feminism—History. 2. Women's rights—History. 3. Male
feminists—History. I. Title.
HQ1154.S94 1999
305.42'09—dc21 98–50838
CIP

This book is printed on paper suitable for recycling and made from fully managed and sustained forest sources.

10 9 8 7 6 5 4 3 2 1
08 07 06 05 04 03 02 01 00 99

Printed and bound in Great Britain by
Antony Rowe Ltd, Chippenham, Wiltshire

Contents

List of Plates

Acknowledgements: Plates 1, 2, 7, 13, 15 and 16 courtesy of the National Portrait Gallery; Plate 8 courtesy of Royal Holloway College Archives; Plate 9 courtesy of the Salvation Army Heritage Centre; Plate 10 courtesy of Canon Christopher Hall.

Every effort has been made to trace all the copyright-holders, but if any have been inadvertently overlooked the publishers will be pleased to make the necessary arrangement at the first opportunity.

Preface and Acknowledgements

In investigating the role of men in the emancipation of women in the late nineteenth and early twentieth centuries a variety of contemporary sources have been used. There are the memoirs, autobiographies, biographies and private papers of people who took part in the controversies or observed them, though often this question was only one of their many concerns. Hansard's records of Parliamentary Debates on the subject over a period of some fifty years are invaluable; for in the end only Parliament could remove the disabilities of women and, if it was prepared to do so, could overrule the opposition of the universities and professions to their admission. Articles in the press and in the prestigious Victorian monthly and quarterly periodicals, as well as cartoons and caricatures, convey how opinions and positions changed.

Some of the most interesting material is to be found in the proceedings, publications and files of professional bodies and philanthropic societies. Among the latter those of the Salvation Army, which did much by its example to raise the status of women, show that this initiative was not taken without misgivings among its male officers. Whilst the stories of the controversy about entry of women into the universities and medical profession is well known and have even been the subject of television programmes, somewhat inadequate attention has been given to the journals and archives of the Associations of other professions, to a number of which Parliament had entrusted the decision as to who might enter and practise in the profession.

In these accounts the motives of men who favoured and men who opposed the admission of women are often exposed with a remarkable and fascinating frankness. Then there are the writings of all kinds of people who helped to form public opinion on the question, not only politicians, educationists and scientists and sociologists, but novelists, poets and dramatists, such as Meredith, Shaw, Wells, Charles Reade, Browning and Tennyson.

Women's Studies in recent years has become such a popular subject that the output of writing touching on the subject of this book is so voluminous as to make a bibliography an impractical addition to a short study. It is hoped that the reference notes will sufficiently enable readers to investigate particular points in which they are interested. Two books, however, which have dealt broadly with the contribution of men to the emancipation of women deserve special mention: Sylvia Strauss, *Traitors to the Masculine Cause* (London, 1982), and A. V. John and C. Eustance, *The Men's Share* (London, 1997).

I should like to thank all those who have helped me, in particular Dr Janet Howarth of St Hilda's College, Oxford, for reading the draft and much improving it by her comments, and Professor Brian Harrison of Corpus Christi College, Oxford, for encouraging me to write the book. I should also like to thank Mrs Joan Baker, Mrs Valerie Barnish, Dr Catherine Bradley, Professor Judith Brown, Dr Geoffrey Carnell, Dr R. A. Cohen, the Revd Tony Cross, Mr Colin Dagnall, Professor Raewyn Dalziel, Canon Christopher Hall, Dr Christine Hallam, Mrs Joanna Heseltine, Dr Elaine Kaye, Mrs Lucy Starr, Ms Lynne Walker and the Hon. C. M. Woodhouse for advice on particular aspects.

I am grateful too for the kindness of archivists and librarians in the Bodleian Library; the Fawcett Library; Friends House, London; Harris Manchester College, Oxford; Knebworth House; the Law Society; the London Library; the National Portrait Gallery; the Oxford Union Society; the Rochdale Public Library; the Royal Academy of Arts; Royal Holloway College; the Royal Institute of British Architects; and the Salvation Army Heritage Centre.

I wish to thank Dr R. F. Watts for permission to quote her unpublished thesis 'The Unitarian Contribution to Female Education in the 19th Century' (Manchester College, Oxford, 1982); Mrs Valery Rose and Mrs Jocelyn Stockley for their helpful and tactful copy-editing, and Mrs Martha A. Kempton, Mrs Dee Campbell and Mrs Marie Ruiz for their efficient secretarial help.

Finally I express my gratitude to Ann Spokes Symonds, but for whom this book would not have been completed.

Oxford RICHARD SYMONDS

All that education and civilisation are doing to replace the law of force by the law of justice remains merely on the surface as long as the citadel of the enemy, in which half the human race is disqualified, is not attacked.... Women cannot be expected to devote themselves to the emancipation of women until men in considerable numbers are prepared to join with them in the undertaking.

John Stuart Mill, *The Subjection of Women* (1869)

Part One

Sapping the Foundations

1 Introduction: Messages from Men of Letters

Man was the problem of the 18th century. Woman is the problem of the 19th.

Victor Hugo

Much has been written about the women who led movements for their equal status with men in the first phase between the mid-nineteenth century and the achievement of the suffrage in 1918. Not a little of the fascination of the story arises from the contrast between the tactics of those who believed that victory would be won by direct, and even outrageous, attacks on the Citadel of male privilege and those who preferred to work through constitutional and conciliatory means and by private and public persuasion. Thus the respective contributions of the militant Pankhursts and of the law-abiding Millicent Garrett Fawcett in the struggle for the suffrage have often been compared. Millicent's sister, Elizabeth Garrett Anderson, appeared, until becoming an uninhibited Suffragette in her retirement, to cultivate the image of a respectable married lady, conventional in everything except in asserting the right, and demonstrating the competence, of women to practise medicine: her undiplomatic, enormously energetic, contemporary Sophia Jex-Blake meanwhile with her devoted little band of would-be women doctors won both wide sympathy and unpopularity by confrontation with and martyrdom at the hands of bewildered male doctors and brutish medical students. In Oxford Annie Rogers and in Cambridge Emily Davies insisted that women should proceed to degrees through the same courses as men; whilst Bertha Johnson at Oxford and Ann Clough and Eleanor Sidgwick at Cambridge were at first prepared to compromise by accepting separate classes for women which might not appear aimed to lead to degree examinations.

The story of these remarkable women makes lively written history and even television. Yet there is a somewhat neglected element in it. Just as Marx and Engels were prophesying at about the same time that the proletarian revolution would only triumph over capitalism with the aid of a fraction of the bourgeoisie, so John Stuart Mill asserted, and most early women feminists recognised, that the Citadel of male domination could not be breached, either by direct assault or by sapping, unless there were some men inside it who would extend a hand to women. The object of this book is to tell the story of some of them between the middle of the nineteenth century when the feminist movement began and 1918–19 when women obtained the Parliamentary vote and entry into most professions in Britain. Occasionally the story has to be carried further as in the Church of England's slow acceptance of ordination of women. With the rapid improvement of communications in the nineteenth century feminist movements in different countries often influenced each other; whilst this study is primarily concerned with what happened in Britain, from time to time therefore it crosses the Atlantic or moves to countries of the Empire and the continent of Europe.

The position of women in Britain in the mid-nineteenth century was harsh. A married woman could not own property and all her earnings went to her husband who was even allowed to beat her or imprison her in the home. If she left the marital home anyone who sheltered her could be prosecuted. Publication of information on birth control was illegal and pregnancies were so frequent as often to result in early death or permanent ill health. There were very few efficient schools for girls except those of Quakers and of some other dissenters to whose lower classes they were sometimes admitted. Women had no voice in the laws by which they were governed. There was hardly any employment open to middle-class women except as governesses or paid companions. Only a few exceptionally gifted women earned an income as writers, often anonymously or under a male pen name. Even if somehow a woman acquired a sufficient education to enter a profession she was barred from obtaining the qualification which would enable her to practise.

There were of course many civilised men who treated the aspirations of wives and daughters with sympathy. Florence Nightingale's father, for example, who was wealthy enough not to need a profession, by teaching her himself and taking her on extensive tours of Europe gave her an education perhaps as good as, and certainly broader than, a son would have received at Eton. Yet even Florence Nightingale's life, despite her achievements, was one of much frustration, being kept at home when she wanted to pursue a career, refusing the man she would have liked to marry rather than accept the social obligations of a Victorian wife, and even when famous having to work by manipulating men behind the scenes instead of leading in the open.

The nineteenth-century movement for women's liberation was a middle-class one. The industrial revolution though it caused them to work in hard conditions had enabled working-class women to get out of their homes. But as a consequence of their work cheap goods had been manufactured which freed upper- and middle-class women from many of their previous household occupations, whilst servants were plentiful and inexpensive. It was unmarried middle-class women particularly who demanded freedom from idleness, and the opportunities to be educated and to earn their own livings so that there was an alternative to uncongenial marriage or endless dependence on fathers and brothers.

By the middle of the century women were beginning to express their grievances. An early success was achieved when the pamphlets written by the well-connected Caroline Norton, describing how she was denied any access to her children on leaving her brutal husband, caused Parliament to pass the Infant Custody Act of 1839. A group of women in which Barbara Leigh Smith (Bodichon) and Bessie Rayner Parkes were prominent set up an Association for Employment of Women and launched the *English Woman's Journal* in the 1850s in which all sorts of grievances and ambitions were aired. Frances Power Cobbe read a paper to the National Association for the Promotion of Social Science in 1862 on 'University Degrees for Women' which provoked much mirth in the newspapers. Elizabeth Garrett had started her struggle to become a doctor. It was women from this group who organised a petition to Parliament that women should be enfranchised under the Reform Bill of 1867, and persuaded John Stuart Mill to support their case in the Commons. Their ambitions could not be achieved without the help of fathers, brothers and husbands of their class in their capacity as voters and formers of public opinion.

The Reform Act of 1832 had enfranchised the middle class. It was only in 1867 that urban workers and in 1885 that rural workers obtained the Parliamentary vote. The men who influenced or took decisions in politics and in the professions formed a relatively small proportion of the population, including Members of Parliament, the gentry who lived off private incomes, landowners, and business and professional men such as lawyers, doctors, ministers of religion and civil servants. Those who had studied at Oxford and Cambridge, where only Anglicans were allowed to graduate, were particularly influential through their network in Government, Parliament and the higher Civil Service. Nonconformists however, often graduates of the new London University or the Scottish universities, neither of which imposed religious restrictions, were becoming active as advocates of social or political reform. It was not unusual for the decision makers to travel in America, continental Europe and the Empire or to be kept informed of ideas and developments there by correspondence with friends and relatives.

Of the newspapers which they read few except the London *Pall Mall Gazette* under W. T. Stead, the *Manchester Guardian* under C. P. Scott, and the *Scotsman* in Edinburgh were sympathetic to the emancipation of women. *The Times* of London, which was read by most Members of Parliament, generally ridiculed or expressed downright hostility to the idea.

There has never been a time, as Britain moved towards general literacy and before the development of mass-readership newspapers, followed by radio and TV, when the influence of books was so important. The way in which attitudes of the opinion formers and of their wives and daughters towards women's rights may have been consciously or unconsciously affected by the books likely to be found in their drawing rooms or studies is therefore of considerable interest.

The middle of the nineteenth century, when a movement for women's rights was beginning to develop, was the heyday of the novel. The most successful were usually printed first in instalments in a magazine and then as a book in three volumes. The novels were distributed throughout the country, until the railways came, in the carts of circulating libraries such as Mudie of London who might purchase as many as a thousand copies of a new work and whose orders could thus exercise an informal censorship of what was considered suitable family reading.

Everyone was reading the novels of Charles Dickens (1812–70) whose characters were so vivid that their sayings often passed rapidly into the English language. Behind the gripping stories was usually a social message; the cruelties in workhouses and schools, the ruinous expense and delay of the law, and the lack of imagination – of freedom to exercise what he called 'fancy' – in the policies of laissez faire, were exposed and ridiculed. Yet though there is warm sympathy throughout his works for the sufferings of women and girls, there is little for their aspirations.

In *Bleak House* (1853), women philanthropists are mocked. Mrs Jellaby holds 'ramification meetings' to organise emigration of the London poor to West Africa. When this project fails because the King of Borriobula Gha wants to sell the immigrants for rum, Mrs Jellaby takes up the rights of women to sit in Parliament. She neglects her children. The family eat at all hours. When they do, she sits at the head of the table whilst her husband remains quite silent except to whisper to his daughter, 'never have a mission, dear child'. Wonderful things tumble out of her cupboards – bits of mouldy pie, odd boots and shoes, saucepans, books with butter sticking on the binding, tails of shrimps and her bonnets. Another member of the Ramification Committee, the grim Mrs Whisk, tells its meetings that its object must be emancipation from the thraldom of the Tyrant Man, who is seeking to confine woman to the narrow sphere of the home. At a wedding she treats the proceedings as part of woman's wrongs. Her colleague, Mrs Pardiggle, pounces on the poor and applies benevolence to them like instant waistcoats.

In 1851 in *Household Words*, the journal which he edited, Dickens used the introduction into Britain by 'Mrs Colonel Bloomer' of a new female nether garment to attack the whole idea of women's emancipation and of American attitudes to it. Should we love a wife or daughter better, he asked rhetorically, if she were a member of Parliament, a parochial guardian, a grand juror or distinguished for her able conduct in the chair? Did we not on the contrary rather seek in her society a haven of refuge after the considerable bow-wow out of doors? He lamented that Mrs Fry had not been content to do good work in jails and that Grace Darling had not gone back to her lighthouse after her noble shipwreck rescue, but that instead these ladies 'went up and down on the earth requiring all women to come forward'. In general he deplored the noisy way in which women agitated at their mission. As for Mrs Bloomer, he observed that America was not generally known for its domestic rest.

Yet Dickens' warm heart led him to take up vigorously one cause, the exploitation of unmarried middle-class women in the main profession open to them – that of governess. His own sister was a governess and his mother, though quite unqualified, opened a school for girls when his father was imprisoned for debt. He supported the demand for girls to be allowed to study at the Royal Academy's art school and he sent his daughter to Bedford College. His novels are full of descriptions of girls' schools which teach a smattering of everything and a knowledge of nothing; but his ideal of a curriculum for them appears mainly to be instruction in the domestic arts, particularly in needlework. He also gave invaluable publicity to Caroline Chisholm in her campaign to improve the treatment of single women who emigrated to Australia, where often they were innocently taken straight off the immigrant ships into brothels. The public's identification of Mrs Jellaby with her, however, harmed Mrs Chisholm's cause and underlined Dickens' ambivalence towards able and independent women.[1]

The entertaining novels of Dickens' prolific contemporary Anthony Trollope (1815–82) were also unobjectionable family reading, though occasional indelicacies had to be removed to appease Mudie's Circulating Library. Anthony Trollope was fascinated by the balance of power within marriage in the upper and middle classes and by the situation of their unmarried daughters. He understood the despairing boredom of the latter; but he saw this as a problem of unused energy, to be solved by marriage to a worthwhile man and then by wielding power behind the scenes.

Later, when he came into the circle of Emily Davies, Elizabeth Garrett and other early feminists, independent women began to appear in his books. He was worried, however, by the feminists' resistance to marriage and horrified by the idea of competition between the sexes and of their assimilation in political privileges and education. In one of his later novels, *Is He Popinjoy?* (1878), there is a grotesque account of a public meeting at the Rights of Women Institute Established for the Relief of Disabilities of Females, known for short as 'The Disabilities'. One speaker is the Baroness Banmann (a very stout woman with a considerable moustache) whose refrain is 'de manifest inferiority of de tyrant saix'. She was modelled on the German Baroness Bulow and the British suffragist leader Lydia Becker, who was of German origin. The other is the pretty but studiously severe young American, Dr Olivia Q. Fleabody, who wears trousers and whose glasses shine disagreeably. She seems to have been drawn from an early woman doctor, Dr Elizabeth Peabody, whom Trollope had met in Boston, and from Elizabeth Blackwell, the first woman doctor to practise in Britain, who had been trained in America.

Trollope's novels are full of attractive intelligent women, but when he was old he wrote to a friend that he had only two important convictions, the eternity of the soul and the supremacy of man over woman. Just once he aired his views on the position of women directly, instead of through a novel, in a lecture to a female audience on the 'Higher Education of Women' in 1868. He compared men who were demanding rights for women with the friendly bear which in an effort to brush flies off the face of a sleeping man unfortunately knocked his brains out. 'Emancipated darlings', he concluded, were now neglecting their domestic duties of helping their mothers and comforting their fathers.[2]

George Meredith (1828–1909), whose writing career extended over fifty years, was the first popular novelist to write with sympathy about intellectual middle-class women struggling against convention, although his own first wife had run away from him with another man. As early as 1855 he expressed shame that because of men women could not walk or travel alone and had to wear intolerable clothes. All his novels dealt with the relations between the sexes, which he saw as warfare mitigated by education. He believed strongly in co-education, writing that 'the devilry between the sexes begins at their separation. They are foreigners when they meet and their alliances are not always binding.'

His message was that women should not just feel but think. Their desire for freedom should be encouraged for it was linked with a desire to serve. A girl should come into marriage saying 'we are equals, I in my way, you in yours, the stronger for being equal'. He deplored the constraint put upon women's natural aptitude and powers, describing middle-class women as unable to be spontaneous or to act independently if they wished to be admired; they had to remain on shelves like other marketable wares, avoiding any motion which risked shattering or tarnishing them. His heroines are mostly caged in unhappy marriages from which they escape. In his most famous novel, *Diana of the Crossways*, modelled on the career of Caroline Norton, his heroine Diana enters into a loveless marriage for want of an alternative but leaves her husband in order to lead an independent life as a novelist. There is a happy ending with a new husband. For although Meredith saw social service as a duty and fulfilment for middle-class women, this was always within marriage. No contented single women leading useful lives appear in his novels.

As a young man Meredith was inspired by Mazzini and by the revolutions of 1848 in Europe. Out of this experience came a faith in freedom. He saw the oppression of women by men as similar to the subjection of one nation by another. 'I began to reflect,' he wrote. 'Ever since I have been oppressed by the injustice done to women, the constraint put upon their natural aptitudes and their faculties; generally much to the degradation of the race.' Strength and courage were what women needed; timidity was their undoing; it was better for them to sin than to drift. In his old age he became a warm supporter of women's suffrage, even sympathising with the militant Suffragettes, because they were overcoming their timidity. 'We women,' wrote one of his biographers not long after his death, 'cannot be grateful enough to George Meredith for the position he took up in his writings with regard to women. He filled us with hope.'[3]

Alongside those of Meredith might be shelved the earlier novels of his contemporary Thomas Hardy (1840–1928), which, though gloomy, were not apparently subversive of the moral and social order. Hardy's last important novel, *Jude the Obscure* (1895), however, may well have been banned from the family library, for in it the hero leaves his wife and lives with his emancipated atheist cousin. After his divorce the latter refuses to marry him though she bears him two children. Jude's son by his first marriage hangs the other two children and then himself. Jude perishes from over-exposure. The book was received by critics with horror, and Hardy's wife tried to persuade W. H. Smith to refuse to sell it.

'Humanity as envisioned by Mr Hardy', said a typical review, 'is largely compounded of hoggishness and hysteria.' Hardy considered that 'whilst civilisation had been able to cover itself with glory in the arts, religion and science, it had failed to create a satisfactory scheme for the conjunction of the sexes'. Throughout his novels runs the theme of the injustice of a social system which thrusts upon women the burden of sexual responsibility and guilt. In contrast to Meredith his profound sympathy is not for middle-class but working-class women. After *Jude*, and another novel which was already in the pipeline, he only wrote poems. In 1906 he told Millicent Fawcett, 'I am in favour of women's suffrage because I think the tendency of the woman's vote will be to break up the present pernicious conventions in respect of women, customs, illegitimacy, the stereotype of the household (that it must be the "unit of society"), the father of a woman's child that it is anybody's business but the woman's own, which I got into hot water for touching on many years ago.' It was a point of view shocking to his contemporaries, hardly acceptable even by the next generation but enthusiastically embraced by the generation after that.[4]

Whilst Hardy was driven into silence, the plays of Shaw, whose message if digested was equally subversive of the social order, failed to scandalise because it was generally regarded as the teasing of an Irish jester with his tongue in cheek. At the end of the nineteenth century and in the first decade of the twentieth George Bernard Shaw (1856–1950) was at the height of his fame. His published plays with their brilliantly readable and provocative prefaces reached as many influential homes as almost any novel. Whilst some other prominent Fabian Socialists, such as Sidney and Beatrice Webb, were dubious about the emancipation of women, for Shaw it was a major preoccupation. His startling message was that woman must become 'unwomanly', even if this meant subordinating her family duties to her emancipation.

His plays were insidious because they were so amusing. Shaw indeed regarded art as the most effective form of propaganda. *Mrs Warren's Profession* was banned from public production as immoral and a later play, *Press Cuttings*, as libellous of recognisable politicians who opposed women's suffrage: but the ban increased their readership. His feminism was an integral part of his socialism. In *Mrs Warren's Profession* the upper-class Madam is shown as no worse than either the City Councillors who were her landlords or the employers of girls in match factories whose starvation wages and dangerous conditions caused them to find prostitution a more attractive occupation. A woman with illegitimate children, Shaw argued, was better off than without them because the father was legally obliged to support her. The only remedy against the white slave trade was a minimum wage law.

In *Major Barbara* the heroine realistically leaves the Salvation Army to marry the heir to her father's armaments firm and thus acquire more power to improve the lives of the poor than she can achieve within a Christian charity. *Pygmalion* reflects not only Shaw's conviction of the potential of education for women, but once more a heroine's realistic attitude to power. Eliza the flower girl, when her acquired accent turns her into a lady, does not as expected marry her autocratic teacher Higgins but prefers, as the author made clear in a postscript, the weak and amiable Freddie whom she can dominate.

Above all Shaw appealed for imagination in attitudes to the capacity of women. 'If we have come to think that the nursery and the kitchen are the natural sphere of women,' he wrote, 'we have done so exactly as English children come to think that a cage is the natural sphere of a parrot because they have never seen one anywhere else.' Domestic slavery, he showed, could be as stultifying as industrial slavery. Prudery was another of his targets, taking as an example the banning as immoral of a film which showed how young girls were lured to London to become prostitutes; the censors had thus become allies of the white slave traffickers in preventing their exposure. He was not content to be an artist and propagandist. He signed petitions for women's suffrage, whose main object he considered should be to get women into Parliament and improve its quality. He petitioned for the release of Suffragettes in prison. He wore out the Inland Revenue authorities by insisting that his wife should be separately taxed because he had no moral right to ascertain her income.

His writings had a great appeal for the kind of middle-class women who were by the end of the nineteenth century admitted to universities and employed as teachers but still barred from most professions, subject to unequal marriage laws and without the Parliamentary vote. Because his plays were as entertaining as they were didactic they also influenced intellectual men of the governing classes for whom attendance at their First Nights became an important social event. To them his unwomanly women, bright and self-possessed, overtly or subtly in conflict with society, often appeared more exciting and attractive than conventional womanly women. Through the ideas which were infiltrated in his plays and other writings many men came to sympathise with a programme of further emancipation which would allow wives, sisters and daughters to evolve as independent individuals with the same political and social rights as themselves.

Usually Shaw was amusing and provocative. Occasionally, however, he declared his convictions with stark earnestness, notably in *The Quintessence of Ibsenism* in 1896. In this he wrote that women were treated as a means to an end, and that this was to deny a person the right to live. 'A whole basketful of wealth of the most sacred quality', he said, 'will be unearthed by the achievement of equality for women and men.'[5]

It was another Fabian Socialist, H. G. Wells, whose novel *Ann Veronica* caused such scandal that daughters may have had to read it under the bedclothes. Published in 1909 and rapidly republished with huge sales, its heroine runs away from her dull suburban family to study biology at London University. She takes part in a Suffragette raid on Parliament and is briefly imprisoned. She next falls in love with her biology instructor, an unhappily married man, and persuades him to elope with her to Switzerland. Eventually they are able to marry and live happily ever after. The book was violently attacked in the press, not only because the heroine's sinful conduct was rewarded but because she and her much older lover were clearly modelled on Wells himself and Amber Reeves, the daughter of one of his Fabian colleagues, who had a child by him. Wells indeed practised free love as enthusiastically as he wrote about it; three women other than his wife became pregnant by him within four years about this time. In his other works he included women in the elite class who should rule the world through an international government, but it was *Ann Veronica* which influenced advanced middle-class young women to abandon conventions and inhibitions more than any other novel of its time.[6]

As well as novels with messages, there were likely to be found in nineteenth-century drawing rooms and studies books by two popular prophets, Carlyle and Ruskin. Thomas Carlyle (1795–1881), though married to an outstandingly intelligent wife, had little sympathy for the aspirations of women in his cult of the hero. Only in his description of Madame Roland in *The French Revolution* does admiration for a woman in political action reluctantly make itself felt. In general he laid down that 'man should bear rule in the house and not the woman. This is an eternal maxim – the law of nature herself – from which no mortal departs unpunished.'[7]

John Ruskin (1819–1900) in his early phase greatly influenced taste in art and architecture; later he wrote more on economics and social questions, both inspiring and exasperating his many readers. His attitude to women was chivalrous but hardly progressive. He frequently attacked the philosophy of 'that poor cretinous John Stuart Mill in endeavouring to open other careers to English women than that of the wife and mother'. Woman, he believed, 'had the power not for invention or creation but sweet ordering'. Her function was to praise. Her education should be aimed at giving her the knowledge to understand and aid the work of man; she should read history but not theology and know languages and science only so far as might enable her to sympathise in her husband's pleasure and in those of his friends. In short, 'the aim of all right education for a woman is to make her love her home better than any other place; that she should as seldom leave it as a queen her queendom; nor ever feel at rest but within its threshold'.

Whilst there was no support for feminists in his writings, in practice Ruskin made a notable contribution to opening up a new career for women; he financed Octavia Hill's scheme for improving the housing of the poor, in which all the visitors employed were women. He befriended many girls, teaching them drawing and botany, obtaining their help in copying out passages of books and drawings and giving talks to girls' schools on religion, mineralogy and mythology. Paradoxically, though in principle he maintained that the place of women was the home, the rather sentimental but thorough training which he gave to these girls not seldom became the basis for their independent careers. Also, like Dickens, he was a champion of governesses – 'You let your servants treat governesses', he protested, 'with less respect than housekeepers, as if the soul of your child were less than jars and groceries.'[8]

Victorian writers on science also exercised a wide influence. Charles Darwin himself lived in retirement, unsolicitous of further publicity after the attacks on his theory of evolution by religious leaders. After him the most popular scientific writer was his disciple T. H. Huxley (1825–95), who had no dislike of controversy. In 1860 he wrote to his colleague Charles Lyell that he 'did not see how we are to make progress while one half of the human race is sunk, as nine tenths of women are, in mere ignorant, parsonese superstition. Five sixths of women will stop in the doll stage of evolution to be the stronghold of parsondom, the drag of civilisation, the degradation of every important pursuit with which they mix themselves, "intrigues" in politics and "friponnes" (cheats) in science. I intend to give my daughters the same training in natural science as their brothers.'

In one of his 'Lay Sermons' in 1865 he attacked the absurd system under which women were educated either to be drudges or toys beneath man or angels above him and were taught that independence was unladylike and blind faith the right state of mind. He called for the recognition that the mind of the average girl is less different from that of the average boy than one boy is from another. Women should be given the same civil and political rights as men. Let them become merchants, barristers, politicians, with a fair field and no favour and they would find their own place – 'because potential motherhood is her lot, woman will be frightfully weighted in the race of life. The duty of man is that not a grain is imposed on that load beyond what nature imposes; that injustice is not added to inequality.' He championed the admission of women to universities and medical schools and was an inspiring teacher to them. Whilst he maintained that women should be treated as equals in public life and in scientific institutions, he considered that this would only be fully possible however when they were as well educated as men and when the disappearance of Victorian prudery allowed free discussion between the sexes.[9]

Among the readers of poetry there was still a cult of the early nineteenth-century Romantics. Of these only Shelley had been a feminist. However much admiration there was for his poetry there was little for the political and social message of this republican atheist who preached free love. The two most influential poets of the Victorian age, Tennyson and Browning, had very different views on the nature and potential of women. For Alfred Lord Tennyson, his son Lionel recollected, the two great social questions in England were firstly the housing and education of the poor man before making him master and secondly the higher education of women. Even when he was an undergraduate at Trinity College, Cambridge, the Apostles Society, of which he was a member, had taken up the question of women's education. Through his dearest friend Arthur Hallam, at that time he drew from Dante a high ideal of woman as an instrument for conveying God's love to man. But when it came to the objects of higher education he did not see it as providing equality of opportunity but rather as restoring an old moral tradition in place of the showy accomplishments provided for contemporary upper and middle-class girls in order to enhance their attractiveness on the marriage market.

Tennyson established his reputation as a popular poet with his long poem *The Princess*, published in 1847. This was a kind of fairy story to which members of a group of young upper-class men and women contributed consecutive episodes on a picnic in the grounds of a stately home. In it Princess Ida founds a university for women, guarded by Amazons, with a death penalty for male intruders. A prince from a neighbouring kingdom, to whom the princess has been betrothed as a child, makes his way into the university with two companions and is discovered and imprisoned. His father brings to his rescue a force which is defeated by the princess's brothers in a tournament. The prince and his companions are wounded. The university then becomes a hospital. In nursing them the princess and her companions fall in love with the victims. Their maternal feelings are aroused by a wandering child. The convalescent prince rescues the princess when she falls off a bridge. Everyone pairs off in marriage and the university project is abandoned.

To this preposterous story Tennyson appends philosophical reflections:

> Either sex alone
> Is half itself and in true marriage lies
> Nor equal nor unequal. Each fulfils
> Defect in each and always thought in thought
> Purpose in purpose, will in will, they grow
> The single pure and perfect animal...

Although *The Princess* was written before Darwin's major works appeared, Tennyson was influenced by the writings on evolution of Lyell, Chambers and others who maintained that the race will regress if there is a failure of physical or mental development in the mother. *The Princess* ridicules the idea that woman only needs the same education as man to achieve her potential. It is a plea for such training as will fit her to perform her own work in the world. Its success however, contributed, as Tennyson's friend F. D. Maurice acknowledged, to the increased interest in female education which caused Queen's College and other early colleges for girls to be established. Yet, as Jane Carlyle noted, Tennyson 'entertained toward women a feeling of almost adoration and ineffable contempt'. The women in the various episodes of his later *Idylls of the King* are remarkably dull except for the few who are unpleasantly wicked.[10]

Whilst Tennyson regarded women as weak, and needing to be tutored, guided and improved by men, there was no trace of patronising in Browning's poems about them. His wife, Elizabeth Barrett Browning, was after all a genius, considered as a possible Poet Laureate whilst he was still unappreciated. Her *Aurora Leigh* impressed on him the claim for free development of women as well as their distinct contribution to the progress of the world. In his poems women are recognised as equals in friendship and often more perceptive in seeing truth and impressing it on men. He did not exalt the characters of the wide variety of women whom he portrayed above those of their male counterparts but was constantly attracted by the intensity of their lives. He described the selfishness of men who forget that women have their own individuality to develop. His alleged obscurity, which irritated reviewers, presented few problems for younger readers among whom in his old age Browning Societies grew up, notably in women's colleges who entertained him to tea and sat at his feet.[11]

This study is divided into two parts. The first is concerned with a preparatory stage in which men were assisting women to lay the foundations of, and acquire the experience necessary for, a successful assault on the Citadel of male privilege. It was John Stuart Mill who by carrying the question of women's suffrage to a vote in the Commons and by writing *The Subjection of Women* first placed the broad question of women's rights before a wide public. There had been earlier champions among the Whigs. Henry, Lord Brougham (1778–1868), in 1850 had introduced an Act which stated that all legislation importing the masculine order should be deemed to include females unless the contrary be provided. This, however, hardly proved to be more than a warning to future draughtsmen. A few years later he introduced a Married Women's Property Bill in the House of Lords. Equally notable was his insistence that women should be admitted as members of the National Association for the Promotion of Social Science, which he founded in 1857 and presided over for its first five years. The association became the most important forum for the early women feminist leaders. Another early Whig sympathiser had been Sir Thomas Talfourd, a friend of Wordsworth, Coleridge and Lamb, who had been inspired by Caroline Norton's misfortunes in 1831 to pilot through Parliament the Infant Custody Bill, which allowed women access to their children when separated from their husbands. There had even been asides in speeches, by Disraeli among others, to the effect that some women might deserve to be given a vote sometime. Mill, however, was the earliest MP to move a resolution that women should be enfranchised on the same basis as men and also the first man to write a book about their disabilities in their entirety and thus to provide a philosophic social and political basis for the emancipation movement.

Whilst Mill had provided the movement with a textbook and a Parliamentary precedent, it was in Josephine Butler's campaigns against the Contagious Diseases Acts and child prostitution that women gained the practical experience which was needed in their future causes. Three men – her husband Canon George Butler, the politician Sir James Stansfeld, and the journalist W. T. Stead – were essential allies in these campaigns, each suffering for his participation. Before women could aspire to regular careers in public life and in the professions, they needed to obtain control over their own bodies. Although the birth control movement was led in its later stages by women, notably Margaret Sanger and Marie Stopes, it was pioneered by male radicals and by a few courageous doctors, of whom several were imprisoned or barred from practising. Later it was the King's physician, Lord Dawson of Penn, who made a devastating attack on the Church of England's opposition to birth control and led his own profession to accept and advise on it. If the nineteenth century, as Victor Hugo said, was the century of the problems of women, the twentieth is likely to be remembered as that of the end of empires and in which brown and black races throughout Asia and Africa achieved independence. In India, which was Britain's largest possession, it was Gandhi who brought women out of their homes into the freedom struggle, insisting that this was part of their wider fight for emancipation. They were rewarded for their participation by being given equal rights with men in the Constitution of the Indian Republic.

In the second part of the book the assault on the Citadel is seen to be launched. Whilst the long fight for the suffrage continued, on another front the earliest objective had to be provision of secondary education and acceptance into universities, without which entry into the higher ranks of the professions and public service was impossible. A number of men with various motivations helped to bring this about. F. D. Maurice was inspired by Christian Socialism when he established Queen's College, London, as a secondary school which became a nursery of women leaders in many spheres. George Grote and Henry Morley regarded the opening of London University to women as part of its radical mission to reach a much larger proportion of the population than that catered for by the still mediaeval and clerical Oxford and Cambridge. At Cambridge, Henry Sidgwick led the movement for entry of women partly because he believed that this would lead to a modernisation of the curriculum for men also. The wealthy pill manufacturer Thomas Holloway was moved to create an early and magnificent women's college, not only by philanthropy but by a somewhat vain desire to 'beat into fits' the college which had been founded by the American businessman Matthew Vassar.

Once there was a basic educational structure, the most obvious profession for women to seek to enter was that of medicine, in which there was a considerable demand for the services of female physicians from middle-class women who in a prudish era disliked being attended by men. Their entry was fiercely opposed by a conservative profession whose members hardly disguised their fear of financial loss through competition from women. It was essential that a few doctors should champion their cause within the bodies which regulated the profession.

Victory, however, was as much due to journalists, writers and Members of Parliament who took up the cause in disgust at the trade unionism of the doctors and were sometimes prodded into action by their wives and daughters. It was Parliament too which established the registration and training of midwives, exasperated by the long and effective opposition of general practitioners who feared that availability of trained midwives would diminish their own earnings.

The manner in which the position of women was treated by churches and religious denominations had a wide significance in an age in which their teaching and tenets carried more weight than in those which preceded and followed it. The Quakers from their origins treated women as the spiritual equals of men. Unitarians were even more active in most of the campaigns for women's emancipation. It was, however, William Booth's action in placing women in leadership positions in the Salvation Army, often above men, which had the most startling impact on attitudes to women's role in religion. The position of the Church of England as the National Church had a wider influence than that on its own members. It only came to accept ordination of women as priests after a very long debate in which, to the general surprise, victory was stimulated by the action of two innovative and tenacious Bishops of Hong Kong.

Mill had expected that women would win the Parliamentary suffrage within his lifetime; once this was gained he was confident that the barriers to their entry into the professions and public services would immediately come down and that social disabilities would be removed. In fact, women obtained entry into the universities and the medical professions and gained some important social rights such as that to own property, before they won the vote. In the fifty-year campaign for the suffrage victory was long delayed because whilst the Liberal Party was generally favourable most of its successive leaders were strongly opposed and in the Conservative Party the position was exactly the reverse. In the long controversy two groups within the Liberal Party were, from the first, active supporters of women's suffrage. One was of intellectual followers of Mill who mostly had strong Cambridge connections. In the other were Radical Nonconformists, often industrialists from the North and Midlands. Among Conservatives the sympathisers were mainly aristocrats, used to associating with intelligent women of their families or class, who were shocked that the latter's grooms and gardeners could vote whilst they could not. Only the Labour Party was fully committed to women's suffrage, which was felt by its leaders, many of whom were active Nonconformists, to be an integral part of socialism, though some trade unionist members remained unenthusiastic. Mill's prediction proved correct in one respect. Very shortly after women gained the suffrage the Sex Disqualification (Removal) Act 1919 enabled women to enter the remaining professions as well as most branches of the higher Civil Service. Yet whilst the bastions of the Citadel had now fallen, there remained inner cells of power from which women were still excluded, in which policies of national importance were informally initiated and appointments to top posts discreetly discussed. Among them were the Gentlemen's Clubs in London and the Senior Common Rooms of those ancient colleges of Oxford and Cambridge which continued to provide the leaders of Government and Civil Service. These were not to be captured until late in the twentieth century. How that came about would be another story.

2 John Stuart Mill Raises the Standard

What in enlightened societies colour, race, religion, or in a conquered country nationality, is to some men, sex is to all women, a peremptory exclusion from almost all honourable occupations.

J. S. Mill, *The Subjection of Women* (1869)

Throughout this book will be found the spirit of John Stuart Mill (1806–73) inspiring not only those who worked for women's suffrage but those who insisted that women should be educated and enabled to have their intellectual capacities tested in competition with men and thus to enter the professions. He was recognised in his own time and since, throughout the English-speaking world and even beyond it, as the pioneer of the movement for equal rights for women. He had many other interests as an economist, social reformer, administrator and Member of Parliament. His attitude to and involvement in the woman question was shaped first by his unusual educational background and then by his relationship with Harriet Taylor. Both were described in his moving *Autobiography*.

His father James Mill was a leading Utilitarian and close friend of Jeremy Bentham and earned his living as Chief Examiner (the senior administrator) in the office of the East India Company (EIC) in London. John was educated entirely at home by his father except for spending a year with a family in France. He was brought up as, and remained, a religious agnostic. He was thus quite free from the conventional influences by which his contemporaries in public life had mostly been moulded, those of the Church, Public School and University. Taught across a table at which his father was writing his influential and prejudiced history of India, John learnt Greek and mathematics from the age of three and became a child prodigy in the classics, logic and history. At eight he taught Latin to his younger sister and at thirteen was taken through a course of political economy. Above all, his father taught him to think for himself rather than to accept what was written. His amusements were in experimental science and he was kept from the company of boys outside the family who might teach him to waste his time.

When he was seventeen his father obtained an appointment for him in the London office of the East India Company where he rapidly rose to earn a satisfactory salary and eventually a generous pension when the EIC was abolished. He was thus freed from financial cares. He got through his official work in three or four hours a day and had ample time for his other interests, including the informal editorship of the *Westminster Review*, which became the most important organ of the Benthamite Radicals. He became, as he described himself retrospectively, a 'reasoning machine', almost wholly formed by his father's philosophy of contempt for and suspicion of any cultivation of the feelings.

When he was twenty, however, he had a breakdown. He emerged from it with a reaction against the arid teaching of his father and Bentham, immersing himself in music and poetry, particularly that of Wordsworth, and in the idealist socialism of St Simon. From then on he decided that the only people who are happy are those whose minds are fixed on the happiness of others. It was now that he met Harriet Taylor, a married woman with two children, whose husband was a well-to-do partner in a firm of wholesale druggists. Mill and Harriet had many common interests and enthusiasms. He started by teaching her; gradually she became the teacher; rapidly they fell in love.

Mill's descriptions of Harriet Taylor in his *Autobiography* and his introduction to her essay on 'The Enfranchisement of Women', both published after her death, abound in superlatives which his friends found embarrassing. She had, he said,

> a mind and heart in which the rarest and what are deemed the most conflicting excellences are unparalleled in any human being I have known or read of.... All that excites admiration when found separately in others seemed brought together in her.... So elevated was the general level of her faculties that the highest poetry, philosophy or art seemed trivial by the side of her.... I venture to prophesy that if mankind continue to improve, their spiritual history for ages to come will be the progressive working out of her thoughts and realization of her conceptions.[1]

J. A. Roebuck described Mill at the time he met Harriet as 'although possessed of much learning and thoroughly acquainted with the political world, utterly ignorant of what is called society, and above all of women'.[2] As for Harriet, a female admirer remembered her as 'possessed of a beauty and grace quite unique of their kind, with a swanlike throat, large dark eyes, with a look of quiet command in them, and movements of undulating grace'.[3] Even Carlyle, who came to dislike her and deplore her effect on Mill, described her as having a 'will of the wisp iridescence' while her husband John Taylor appeared to him 'a dull good man, obtuse and most joyous natured'.[4]

Harriet and her husband came from Unitarian families, and Harriet had ambitions to become a writer like Harriet Martineau. She declared that 'the desire to give and receive feeling is almost the whole of my character'.[5] At first she and Mill made their rendezvous at the Zoo 'in the lover's alley in front of the rhino'. Then they came to behave with great indiscretion, travelling in France and Italy together alone, or later with Harriet's youngest child Helen. Their close friends supposed their relationship to be brazen but innocent: Carlyle called Harriet 'Mill's Platonica'.[6] Her husband did not want a break and provided her with a suburban cottage where Mill could visit her at weekends. After being received in society with what Mill regarded as insults they withdrew from it almost entirely. Most of Mill's contemporaries and many subsequent writers treated his descriptions of Harriet's qualities and her responsibility for his writings as absurd and besotted. Only when their correspondence was published a hundred years after Harriet's death did it become surprisingly clear that her influence over Mill was quite as great as he asserted and that she strengthened not so much the sentimental as the rational element in his thought.[7]

Mill was a feminist before he met Harriet. He had openly disagreed with his father's contention that women did not need the vote because their interests were safeguarded by their male relatives. On his way to the office at the age of 17 he had come across the body of a strangled baby. Shocked by this evidence of infanticide, he accompanied a friend who was distributing birth-control tracts by Francis Place. They were arrested and brought up before the Lord Mayor who sentenced them to two weeks' imprisonment but released them after a day or two.[8] Mill's generous attitude to the situation of women was one of his first attractions for Harriet, who had recently written a paper on marriage, which she described as

> the only contract ever heard of, of which a necessary condition in the contracting parties was that one should be entirely ignorant of the nature and terms of the contract. For owing to the voting of chastity as the greatest virtue of women, the fact that a woman knew what she undertook would be considered a just reason for preventing her undertaking it.[9]

On all Mill's work written after their meeting Harriet can now be seen to have had a major influence both through their discussion and in her comments on Mill's drafts. This was particularly striking in his *Principles of Political Economy*. Harriet's passionate sympathy with the revolutions throughout Europe in 1848 brought about such substantial changes that whilst the first edition had been a favourable exposition of laissez faire economics, the second edition was so sympathetic to socialism that it became a basic text in the evolution for British socialists. In the third edition Mill also incorporated passages proposed by Harriet on the disabilities of women.[10]

In 1849 John Taylor became fatally ill with cancer. Full of remorse, Harriet hurried back from a continental holiday with Mill to nurse him. Eighteen months after his death Mill married Harriet in a registry office after drawing up a document renouncing all legal rights over her as a husband. Mill and Harriet now realised that they were both suffering from tuberculosis. Not knowing how much time was left, they drew up plans for future writing which Mill was to follow faithfully after Harriet's death. Except for his *Logic* the principles underlying his remaining important works were defined though not actually composed by Harriet. The only piece written by Harriet herself after their marriage was published anonymously in 1857 in the *Westminster Review* and later revealed by Mill as her work. This essay on 'The Enfranchisement of Women' was to provide the basis for much of Mill's great book on *The Subjection of Women*.

Harriet took as her starting point a recent convention of women in the United States which had made three demands for equality with men. Firstly in education at primary, secondary and university level and in medical, legal and theological institutions; secondly for partnership in the labours and gains, risks and remuneration of productive industry; and thirdly in the formation and administration of laws at the municipal, state and national levels. These claims, she urged, were equally valid in Europe. She went on to argue that the fact that an institution or practice was customary was no proof of its goodness. Freedom of industry, of the press, and of conscience had all been opposed on grounds of custom. 'We deny', she said, 'the right of any portion of the species to decide for another portion what is their "proper sphere".... Let every occupation be open to all, and employments will fall into the hands of those men and women who are found by experience to be most capable of exercising them.' She had no patience with moderate reformers who said that women should be educated to be companions of men in general knowledge, poetry, and even coquetting with science. 'What is wanted for women,' she declared, 'is equal admission to all social privileges, not a position apart as a sort of sentimental priesthood.... We are firmly convinced that the division of mankind into two classes, one born to rule over the other is . . . an unqualified mischief.'[11] Harriet's article was reprinted and widely distributed in America, where she became much better known in her own right than in Britain.

Although their relationship was now respectable, the Mills lived quietly in Blackheath, seven miles from London, seldom venturing into society. In the winter of 1858 they set out for the South of France for their health. On the way Harriet died in a hotel in Avignon and was buried there. Mill immediately purchased a cottage close to the cemetery and furnished it with everything from the room where she had died. There he lived for much of the year with Harriet's daughter Helen, visiting the grave daily, botanising in the mountains and continuing on the programme of writing which had been laid down for him.

This tranquil life was interrupted by election as Liberal Member of Parliament for Westminster. His object in entering Parliament was to promote and draw attention to ideas which were so unpopular that no one else would handle them. Among these were electoral reform, land reforms in Ireland, sanctuary for political refugees, proportional representation, and the need for a London County Council; his keenest interest was in the admission of women to the franchise. When he entered Parliament his party, the Liberals, were in office, with Gladstone as their leader in the House. Whilst Gladstone always listened to him he had no sympathy for the enfranchisement of women.

On 20 May 1867 a formal proposal was made for the first time in the House of Commons to extend the Parliamentary vote to women, when Mill rose, having given previous notice, to speak on the Reform Bill which was before the House and which would give the vote to all males who were householders. He simply proposed that the word 'man' should be replaced in the Bill by 'person'. He started in a low key. He had no desire, he said, to alter the balance between classes; indeed more women householders might be likely to vote Conservative than Liberal. His argument was one of expediency. It could not be claimed that women who managed estates, conducted businesses, or who were heads of families or school mistresses, if admitted to the franchise, would revolutionise the state or cause worse laws to be passed. Their present exclusion violated one of the oldest constitutional maxims, that of no taxation without representation.

He refuted the argument that 'politics are not women's business' by pointing out that they were not the regular business of most men either;

the occupations of most women were indeed domestic but these were compatible with a keen interest in national affairs. The notion of a hard and fast line between men's and women's occupations belonged to a bygone society. Among the educated classes a wife was now her husband's closest companion. Was it good for a man, he asked, to live in communion with one who was studiously kept inferior to himself. Unless women were raised to the level of men, men would be pulled down to theirs. The denial of the vote was a proclamation that society did not expect women to concern themselves with public interests and it shaped the form of girls' education.

Passing to the argument that women were already adequately represented by their husbands, Mill allowed his indignation to take over. He would like, he said, to see statistics of the number of women annually beaten and kicked to death by their male protectors and of the sentences which the latter received, as compared with sentences imposed at the same sessions for unlawful taking of property. 'We should then have an arithmetical estimate of the value set by a male legislature and male tribunals on the murder of a woman.' He pointed out how because they were disenfranchised women suffered in education and entry to the professions. They were unable to enter the universities. Endowments which had been made for the education of all children had been manipulated so that girls were excluded from their benefits. If they did succeed in obtaining an education, the only posts open to them were as governesses. He gave the specific example of Elizabeth Garrett who in her long struggle to enter the medical profession had finally found one door left accidentally open, that of the Society of Apothecaries. This door had been immediately closed after her, so that no other woman could qualify through it. Similarly although women had once been Associates of the Royal Academy they were no longer eligible. As for married women, all that a wife owned belonged in law to her husband who might squander every penny of it. The richer classes managed to avoid this by making marriage settlements for their daughters, but the working wife of a poor man was unable to have any voice in the disposal of her earnings.

In his peroration, prophetically foreseeing the violence of the Suffragettes forty years later, Mill declared that grievances less than these had provoked revolutions.[12] His arguments were powerful and caused some Members to change the way they voted. He was warmly supported by Henry Fawcett, who described the success of the local examinations held at Cambridge which had shown that girls could quite hold their own against boys in Latin, Greek and Mathematics. Mill's speech, he said, would have a great effect in the country for bringing the question of women's suffrage out of ridicule. This was hardly indicated in the following interventions of his opponents. The Liberal Samuel Laing relied on Shakespeare to show that Juliet, Ophelia and Desdemona would not have interested themselves in voting in elections. The only precedent for women taking an active part in public life, he said, was that of the Amazons of Dahomey and this was not one to be emulated. Viscount Galway implored Mill in vain to withdraw his amendment because voting on it would place many gentlemen in an embarrassing position. The familiar argument was also put forward that women would be debased and degraded by participating in elections, a point met by the Liberal Sir George Bowring by suggesting the use of ballot slips.[13]

Mill's amendment was defeated by 196 votes to 73. It was a not discouraging start and among his supporters were men of considerable distinction including John Bright, Tom Hughes, Henry Labouchere and J. E. Gorst. Gladstone voted against. Disraeli abstained; though he had expressed the opinion that women with property should be given the vote, no doubt he found it imprudent to support Mill in light of the general feeling on the question in the party which he led.[14] Later Mill introduced a Bill which eventually, after his departure from Parliament, provided the first step in enabling married women to own property.

In the General Election of 1868 Mill, although refusing to contribute to the expenses of his own campaign, made an unpopular public gesture of support by contributing to those of the atheist Charles Bradlaugh at Northampton. He was soundly defeated by the Conservative candidate, W. H. Smith, who had spent lavishly, and though approached by several constituencies, he never ran again. It has been surmised that if Mill had remained in Parliament women might have obtained the vote much earlier.[15] How and why it took fifty years more for them to do so is discussed in Chapter 9.

After Harriet's death her daughter Helen became Mill's constant companion and collaborator. 'Surely,' wrote Mill in his autobiography, 'no one ever before was so fortunate, as, after such a loss as mine, to draw another prize in the lottery of life – another companion, stimulator, adviser and instructor of the rarest quality. Whoever may think of me and the work I have done must never forget that it is the product not of one intellect and conscience but of three, the least considerable of whom, and above all the least original, is myself.'[16] It was not until 1869 that Mill's book *The Subjection of Women* was published though it had been long in preparation. 'All that is most striking and profound in it', he wrote, 'belongs to my wife, coming from the fund of thought which had been common to us both, by our innumerable conversations and discussions on a topic which filled so large a place in our minds.' His object was 'to show that the legal subordination of one sex to another is wrong in itself and one of the chief hindrances to human improvement'. Mill's thesis in this book is that in the modern world it should not be ordained that to be born a girl instead of a boy, any more than to be born a black instead of white, should decide a person's position in life and prevent her from entering all respectable occupations. He describes with indignation the life to which many women are condemned, forbidden to exercise the practical abilities of which many of them are conscious. A wife can possess no property independently of her husband. She has no rights even over her children. She frequently suffers bodily assaults without protest, for if her husband is prosecuted and goes to prison the family may starve.

Historically he explains how the subjection of women originally arose from their lack of physical strength. Their present position remains a primitive state of slavery in a society in which the law of the strongest has otherwise been abandoned. Whilst human beings are no longer born to their place and chained to it by a feudal system, the disabilities which women suffer, merely from the fact of their birth, are the only examples of their kind in modern legislation. Marriage is the only bondage known to our law; there remain no legal slaves except the mistress of every house. Men want willing slaves and have therefore done everything possible to enslave women's minds. A woman is brought up from childhood in the belief that her ideal of character is the very opposite to that of man – to live for others, to have no life but family affections and to consider that all objects of social ambition can only be found through her husband.

Women's lack of achievement in philosophy, art, science and other fields, he continued, has been due to inadequate education and preparation. Everything a woman does has to be done at odd times. Yet as rulers European queens and princesses have been notably successful; this has been because the ladies of reigning families have been the only women to be allowed the same range of interests and development as men, so that there is no sense of inferiority. Women have also been successful as writers, but in order to be published and read they must be sycophantic to men. In general the mental capacities of women are largely unknown, and cannot be known until they are given equal opportunities in education and are allowed to compete with men on equal terms. What is now called 'the nature of women' is an artificial thing.

The originality of Mill is the argument that subjection of women is harmful to men. The whole of society, he said, is deprived of the potential services of half its members, notably in such professions as medicine, law and Parliament. It is bad for the character of a boy to grow up believing that by right he is superior to half the human race. And it is harmful for a married man to have to live with someone who is a dead weight, his inferior in intelligence, conditioned to have no opinions outside the family and consequently caring only for what will bring in a title, a place or a good marriage. Women cannot by their own efforts alone escape from subjection, for a woman who joins any movement of which her husband disapproves makes herself a martyr. Thus women cannot be expected in any numbers to devote themselves to their emancipation unless a considerable number of men join them in the undertaking.

Mill on *The Subjection of Women* flows naturally from Mill on *Liberty*. What in unenlightened societies, colour, race, religion, or in the case of a conquered country nationality, are to some men, sex is to all women – peremptory exclusion from almost all honourable occupations. All that education and civilisation are doing to replace the law of force by the law of justice remains merely on the surface as long as the citadel of the enemy, in which half the human race is disqualified, is not attacked.[17]

Of all that Mill wrote, *The Subjection of Women* aroused the greatest antagonism. Fitzjames Stephen described it as the strongest and strangest illustration of by far the most mischievous of all the popular feelings of the age. It was not only conservatives who were dismayed. Frederic Harrison, the eminent Positivist and liberal writer, even 30 years later considered the book 'rank moral and social anarchy'. Mill's friend Alexander Bain, in his biography of him, charges him with 'overstraining', and 'postulating a degree of mental equality which does not chime with experience'. He remonstrated with Mill at the time against putting the case more strongly than people generally would be willing to accept. Mill 'replied with much warmth contending that the day for a temporizing policy was past'.[18]

Although Mill had referred to Queen Victoria as an example of the success of women in public life he by no means had her support. The Queen in a private letter said that she was anxious 'to enlist everyone who can speak or write to join in checking this mad wicked folly of "women's rights".... It is a subject which makes the Queen so furious that she cannot contain herself. God created men and women different. Then let them remain in their own position.'[19] The book however was highly successful in its object of stimulating discussion. It rapidly went into further editions in English, French, German and other languages.

Its ideas were in many ways ahead of its time and emerging leaders of the women's movement such as Millicent Fawcett and Barbara Bodichon regarded it almost as a Bible of women's rights.

After Mill left Parliament he devoted much of his time to encouraging women to organise themselves in the movement for the suffrage. He and Helen took the initiative in the formation of the National Society for Women's Suffrage. He was adamant that this must be a single-issue organisation. Although he himself gave evidence to the Royal Commission on the Contagious Diseases Acts, which he strongly opposed, he insisted that the Society should not involve itself in the question. Mill and Helen persuaded two very influential women, Florence Nightingale and Mary Carpenter, to join the suffrage movement in spite of initial reluctance. The Mills (for it is hard to distinguish which of them now drafted Mill's letters) also corresponded enthusiastically with the Universal Suffrage Association in the USA. There were small successes. The municipal franchise was extended to women in 1869 and they were enabled to serve as members of the school boards created in 1870. Emily Davies and Elizabeth Garrett were among those first elected to do so in London.

In May 1873 Mill died suddenly in his house at Avignon. His last words, according to Helen, were, 'You know that I have done my work.' Whether he was addressing Helen or the spirit of Harriet or the Supreme Being whose existence is hinted at in his posthumous *Essays on Religion* may be a matter for speculation. Helen was indignant when a memorial committee was formed in London and it was proposed to bury Mill in Westminster Abbey. The proposal was utterly foolish and unsuitable, she declared, in light of Mill's views. She need not have worried. A critical anonymous obituary in the London *Times* observed 'of late Mr Mill has not come before the world with advantage. When he appeared in public it was to advocate the fanciful rights of women.' Readers were puzzled by a verse by Thomas Moore quoted in the article,

> There are two Mr Mills, to whom those who like reading
> What's vastly unreadable call very clever.
> And whereas Mill senior makes war on good breeding,
> Mill junior makes war on all breeding whatever.[20]

The obituarist was Abraham Hayward QC, whose lifelong hatred of Mill went back to their youth when in a debating society the latter had trounced him 'as a ploughshare goes over a mouse'. Not content with the obituary, he wrote to a number of eminent men, including the Prime Minister, Gladstone, alleging that not only had Mill distributed birth control pamphlets in his youth but he had subsequently advocated artificial checks to population in his writings; also that when he fell in love with a married woman he had written in favour of easy divorce. Gladstone consulted the Duke of Argyll who found nothing in Mill's works to substantiate the allegations and himself subscribed to the memorial fund. Gladstone however thought it prudent to withdraw.[21] So Mill's remains stayed in Avignon, next to those of Harriet; a handsome memorial to him was erected on the London Embankment instead of in the Abbey. Despite Mill's caution in his adult life on the subject of birth control, the controversy and the publicity which it received were his posthumous contribution to the subsequent campaign for family planning whose success was to prove of great importance in the liberation of women. In his will he left £3,000 'to the first university in England to open its degrees to women' and the same sum to endow scholarships for women.

3 The Allies of Josephine Butler

His right honourable friend (Stansfeld) had sacrificed time, peace, money and every other ambition in order to deal with this question. He did not know of any other instance of a man who had so completely severed himself from every object of ambition in order to devote himself to the one question in which he felt a deep interest.

Samuel Whitbread in the House of Commons (20 April 1883)

Josephine Butler conducted two great campaigns in Britain to obtain action by Parliament. The first was against the Contagious Diseases Acts. The second was to raise the age of consent and stop the export of children to brothels abroad. They were important not only for what they achieved in themselves but because for the first time they brought women out in numbers to speak on public platforms, to lobby and to demonstrate. As a Member of Parliament told Josephine Butler, 'we know how to manage any other opposition in the House of Commons, but this is very awkward for us, the revolt of the women'.[1] Among religious leaders it was mainly the Quakers and Salvation Army who collaborated with Josephine Butler; for her causes were not at first regarded as respectable in the Establishment and the Church of England. Her most constant and valuable ally was her husband, George Butler, who had an impeccable position as Principal of an important college and as a priest, later a canon, in the Church of England. Liberal, generous and believing naturally in the equality of women, he supported his wife with complete sympathy, despite possible harm to his career. Josephine Butler needed friends in Parliament who would press for the necessary legislation irrespective of which party was in power or the damage done to their prospects of office by their obstinate championship of unpopular causes. She had supporters on both sides of the House but much the most influential and persistent was the Liberal James Stansfeld who had held Cabinet office, but was excluded from it during his participation in the campaigns. The leaders in the House of Commons, however, for years refused to allocate time to pass legislation to raise the Age of Consent. To persuade them to do so required outside pressure. The London, though not the provincial, press viewed the questions as too indecent to discuss, until, moved by an appeal from Josephine, her third great ally W. T. Stead, the editor of the *Pall Mall Gazette*, so startled and shamed public opinion that Parliament was forced into immediate action.

THE HUSBAND – CANON GEORGE BUTLER (1819–90)

Josephine Butler was the only post-reformation British woman to be commemorated in the Church of England's Alternative Service Book of 1980. The Bishop of Derby persuaded the General Synod to include her because this would honour symbolically all parsons' wives.[2] If his statement inspired an image of a Mrs Butler immersed in church bazaars and choir practices, as a devoted auxiliary to the vicar, it would be far from accurate. It was rather her husband, George, who played the traditional supportive part of a clerical spouse. George Butler came from a family of headmasters and dons. Liberated from the family earnestness as an undergraduate entering Trinity, Cambridge, George, as Josephine put it, 'preferred the company of men who amused themselves to the more exemplary'. His father, considering him to be wasting his time, sternly arranged for him to make a new start at Exeter College, Oxford, where he won the prestigious Hertford Prize, took a first class degree and became a fellow. He also made lifelong friends in Benjamin Jowett, A. P. Stanley, Matthew Arnold and J. A. Froude. On graduating he became a private tutor and then a tutor at Durham University where he played cricket for the county. His father, now Dean of Peterborough, urged him to take Orders. George refused: 'You know that I don't like parsons,' he wrote to Josephine, 'I think all dressing up and official manner are affectations.' He objected to parts of the Church of England services and supposed that the Virgin Mary would have been baffled by the Athanasian Creed.[3]

Shortly before his marriage he returned to Oxford as a public examiner. When they married he was 32 and Josephine 23. Josephine came from a Northumbrian family of reformers and evangelicals. Her father, John Grey, had campaigned for the Great Reform Bill of 1832, introduced by his kinsman Earl Grey, and later for the repeal of the Corn Laws. When Josephine was in her teens she went through a religious crisis, emerging with a strong sense of the living presence of God. 'The things which I believed,' she was to recall, 'I learnt direct from God. I never sought light or guidance from any saint, man or woman.'[4] Such a wife in the mid-nineteenth century would have been unhappy with a conventional authoritarian husband; George however wrote to her while they were engaged, 'I should think it undue presumption in me to suggest anything to you in regard to your life and duties.... I am more content to leave you to walk by yourself in the path you shall choose.' 'If I can help you by my strength and physique,' he wrote later, 'depend on it I will do so. In all other matters I think you are more capable of giving me aid than of borrowing it.'[5] This attitude was maintained throughout their marriage.

In Oxford the Butlers started a hall for unattached students. George, who was generally ahead of his time throughout his career as an educationist, gave the first lectures in the University on geography as well as on the study of art. In 1853 he overcame his scruples and was ordained, having come to believe that the status would help him in his pastoral role with students. In 1857, largely because Josephine's health was affected by the Oxford climate, he accepted the post of Vice Principal of Cheltenham Boys' College. Here, though George disliked argument and heated discussion, he took a lone stand in supporting the North in the American Civil War, writing a trenchant pamphlet entitled 'Does the Bible sanction slavery?' This made the Butlers unpopular with their colleagues and gave them experience of rowing against the stream.

In 1865 George became Principal of Liverpool College. Josephine's main activity at this time was as President of the North of England Council for the Education of Women, with Anne Clough as its Secretary. In 1869 she edited an innovative book on *Women's Work and Women's Culture*.[6] George too was an early advocate of higher education for women. In his chapter in the same book, on 'Education as a Profession for Women', he maintained, like Mill, that only if women had equal educational opportunities with men would experience indicate the professions for which they were suited. Women teachers, he observed, must be adequately paid, unlike in America, where they earned about as much as a washerwoman and less than the janitress who swept out the school. As a practical way to advance, he proposed that girls' schools should be set up parallel with boys' schools, the girls being taught by some of their masters and sharing libraries and examinations. This would also provide 'some countervailing antidote to the spirit of the Public School'.[7]

It was now that Josephine came to take the lead in opposing the extension of the Contagious Diseases Acts. These had been introduced as a temporary measure in towns where there were military or naval establishments, in order to lessen the incidence of venereal diseases in the forces. Under the Acts a special corps of plain-clothes police was established, responsible for compiling lists of licensed prostitutes. Its officers were empowered to arrest and bring before a magistrate any woman whom they had 'good cause' to believe to be a prostitute. A single magistrate could send the woman for compulsory examination and treatment: there was no right of appeal against his decision. The Acts were supported by the majority of the medical profession and of the clergy. Two medical men, however, wrote to Josephine to ask her to take the initiative in opposing this introduction of licensed prostitution. It was only, they said, the ladies of Britain who could make an effective protest against this insult to their sex. Josephine was well known as President of the North of England Council and for her social work among the poorest women of Liverpool; she was a wife and mother, married to a Church of England clergyman who could be relied upon to support her. She had all the qualifications to lead the campaign. To a Grey the Acts represented as much a political as a moral danger, a legislative, bureaucratic and medical tyranny in defiance of Magna Carta, the Bill of Rights and Habeas Corpus. Yet the task appeared difficult, dreadful and disgusting: 'like Jonah,' she recalled, 'I fled from the Lord'. For three months she wrestled with herself whilst appeals continued to pour in. 'I could not bear the thought of making my dear companion a sharer of the pain; yet I saw that we must needs be united in this as in all else. I had tried to arrange to suffer alone, but I could not act alone.' She was aware how much George would suffer, being by nature quiet, intellectual and domestic, disliking controversy. 'I spoke to my husband then of all that had passed in my mind and said, "I feel as if I must go out in the streets, and cry aloud, or my heart will break". He did not pause to ask "what will the world say?", or "is this suitable work for a woman?" He saw only a great wrong and a deep desire to redress that wrong... and above all God whose call (whatever it may be) is man's highest honour to obey.'[8]

Josephine commenced the campaign with a speaking tour in the northern cities, mainly to working-class audiences, rousing interest and stimulating petitions to Parliament. George organised a flying column of speakers who would address meetings when there were Parliamentary elections, a solicitor who would protect victims of the acts, and hospitals and refuges for prostitutes who wished to begin a new life.[9]

Josephine and her colleagues formed a Ladies' National Association for the Repeal of the Contagious Diseases Acts, which put up candidates at Parliamentary by-elections against those who supported the Acts. This led in one case to the defeat of the Liberals through a split in their vote. Personally this was unhappy for the Butlers who were committed Liberals, but George continued to conduct an amiable correspondence with Gladstone in Latin verse. His duties prevented him accompanying Josephine in the ferocious by-election campaigns. He made a courageous intervention, however, in the quarter where it was most needed. At the Annual Church Congress in 1872 he read a paper on 'The Duty of the Church of England in Moral Questions'. In this he tried to recommend that the Church Congress should protest against the immoral legislation. There were howls of protest from some clergymen who, according to Josephine, were 'young men of good families, put into the church from motives not the highest'. The Bishop of Lincoln, who was in the chair, had to ask George to sit down. He was afterwards rebuked by the Archbishop of York, who had been his friend at Oxford. He replied politely but firmly, advising the Archbishop to read the report of the Royal Commission on the Contagious Diseases Acts, which had just been published, and he nailed his flag to the mast by adding 'I desire fervently to see all unjust and class legislation abolished.'[10] He brought out as a pamphlet the speech which he had been unable to complete, containing what a reviewer called a startling and tremendous warning:

> If we constantly take the wrong side, if we are found continually acting in opposition to the conscience of the people, walking in the steps of those, whether baptised or churchmen, kings or Parliaments, who burnt the martyrs, drove out Wesley and Whitefield, taxed the American colonies, upheld slavery, trafficked in Church preferments, withstood the application of our endowment to purposes of general education, tied up land by vexatious laws, connived at drunkenness and made vice easy and professionally safe – then I think the time is not far off when the cry will come from all parts of the United Kingdom against the Church of England, 'Away with it! Why cumbers it the ground?'[11]

By nature George was not a controversialist and he regarded the Church Congress episode with a certain detached amusement. His heart was in his profession as a teacher – 'a blessed office', as he called it. In addition to his work in the College he lectured, to adult, mainly working-class, audiences mostly on the constitutional history of Britain. 'He is never excited and never exaggerates,' said Josephine, who herself was inclined to self-dramatisation. He encouraged her when occasionally she lost heart. Both of them were good linguists and in vacations he would take her off on continental holidays to relax.[12]

For the Principal of a College, the domestic disruptions must have been considerable. Throughout her marriage Josephine would spontaneously take lame ducks – usually poor, and often former prostitutes – into her home. George invariably welcomed the girls, read to them and prayed with them. At the age of 63 George resigned his position as Principal of Liverpool College. Josephine's family had lost much of their money in a bank crash and the Butlers had spent generously on the campaigns. Their financial situation was worrying but eventually Gladstone made George a Canon of Winchester. The Butlers had a house on the Cathedral Close and as he grew older it became restful for him physically as well as spiritually to take his part in the Cathedral services.

In 1883 the long campaign was successful: the Acts were suspended and finally repealed in 1886. Josephine now took the initiative in the formation of an International Federation for the Abolition of Government Regulation of Prostitution and, as its Secretary, spent much of her time abroad. George participated in the Federation's annual conferences. In his interventions he stressed the importance of religious influence on the movement to raise the Age of Consent in Britain, and of its interdenominational character. The Nonconformists, he said, had stood like a wall of stone. Eventually the Anglican clergy had come round, he admitted, though in general they had some class prejudices to get over before they could look fairly at any movement of and for the people.[13]

George Butler died in 1890. His friend Lord Coleridge described him as 'a man more remarkable in himself than in anything he ever did or wrote, an admirable scholar, a fine artist, full of fun, who took a position below his merits without a jealous or repining thought'. It was this modesty which struck many of his friends. James Stansfeld, who worked closely with him for many years in the cause, said that he had played a secondary part as willingly, as enthusiastically, and with as much determination as if he and he alone had been the leader of the movement.[14] His friends felt it a pity that with his talents he had not achieved more than a headmastership and a canonry. Josephine's campaigns, because they disturbed vested interests, had made enemies who spread rumours – wholly unfounded – that it was domestic dissatisfaction which sent Josephine out of her home on her campaigns. He had, in fact, been the perfect husband for her. As W. T. Stead observed, 'had it been otherwise she could not have begun her work, or if she had begun it, it would speedily have ended. He believed in her implicitly and had a joyous placid conviction that she could suffer no harm while her work was in progress. Had he fretted or grieved she would have broken down.' As Josephine herself said, he was able from the first to correct her judgement and soothe her spirit, having the blessed gift of common sense. He needed no conversion to her causes. He was already on the side of the reformers when he met her and, she said, the idea of justice to women and equality of the moral law seemed to have been intuitive with him.[15]

When they first married it was Josephine who had been the assistant, making maps and copying pictures for George's lectures. Even when they first arrived in Liverpool, she was still regarded as the Principal's wife, rather than as someone with important interests of her own. George accepted and delighted in his wife's changing role. He never saw her a menace to his career, for as she said, he saw the Gospel as revolutionary and he was not afraid of revolution.[16] His example was to be an inspiration to the next generation of professional men and women in demonstrating that the equality in marriage which John Stuart Mill had preached was not an absurd ideal.

THE PARLIAMENTARIAN – SIR JAMES STANSFELD (1820–98)

In 1840 a London University undergraduate, James Stansfeld, was present at the great World Anti-Slavery Convention organised in London by the British and Foreign Anti-Slavery Society. There were 500 representatives, and among the Americans were 20 women. After a bitter debate the Convention voted to exclude them, and in protest William Lloyd Garrison, one of the most eminent of the American abolitionists, took his seat with the ladies as a spectator in the gallery.[17] Stansfeld looked back to this as a decisive moment in his life. 'It was that episode, the shame I felt as an Englishman and a man,' he recalled, 'that first turned my thoughts to the position created in England for our mothers, sisters and wives, that made me resolve that all schemes of education, of political reform, should include them as equals.'[18]

Stansfeld came from a Unitarian family in Halifax. His father was a County Court Judge. Excluded from Oxford and Cambridge as a Nonconformist, James studied in the much more liberal atmosphere of University College, London. After graduating he read for the Bar. He now came under the influence of William Ashurst, whose daughter he married. Ashurst was an ardent political reformer and advocate of equality between the sexes. Asked why he was a feminist, he explained how whilst an Undersheriff of London he had seen a girl tried for child murder who had been betrayed by a man, was convicted by men, sentenced by a man and hanged by a man. 'It made me think.'[19] Through Ashurst Stansfeld came to know and admire Mazzini, who was then a refugee in London and sometimes stayed with the Stansfelds.

In 1850 he gave up the Bar and became the proprietor of a brewery, 'with the explicit intention of making just sufficient to live on that I might devote the rest of my life to public objects'.[20] Meanwhile he made speeches on Parliamentary reform and the need for repeal of taxes on knowledge. In 1859 he was elected as a Liberal MP for Halifax, which he represented until his retirement in 1895. He acted with the radical wing of the party led by John Bright. Many of his early speeches were concerned with the cause of Italian liberty as well as that of Polish independence. Garibaldi referred to him as 'the type of English courage, loyalty and constant friend of Italy in her evil days, the champion of the weak and of the oppressed abroad'.[21] He rapidly won a reputation in the House of Commons for what the *Saturday Review* called 'an indescribable charm of voice, manner and earnestness'.[22] He served as Junior Lord of the Admiralty, Under-Secretary for India, and Financial Secretary to the Treasury before entering Gladstone's cabinet as President of the Local Government Board in 1871. Here, over the strong objections of his officials, he appointed for the first time a woman, Mrs Nassau Senior, as a Poor Law Inspector.

He had proved a competent minister by the time the Liberal Government went out of office in 1874. With John Bright and W. E. Forster he was regarded as one of the leaders of the Liberal radical wing and destined for even higher office when the party returned to power. To the general surprise, however, he put aside other political questions to devote himself, inside and outside Parliament, to the repeal of the Contagious Diseases Acts. He became a Vice President of the National Association for the Repeal of the Contagious Diseases Acts and made his position quite clear in an early speech. 'I have made my choice – I have cast in my lot with those men and women... who hitherto have held a hope which too long has seemed forlorn, and never will I desist and they desist from this sacred agitation until these degrading laws are blotted out from the statute book for ever.'[23] *The Times* 'sincerely regretted that a statesman of Mr Stansfeld's eminence should identify himself with such a hysterical crusade in which it is impossible to take part without herding with prurient and cynical fanatics'.[24]

With the Conservatives in office there was little prospect of early repeal. What could be done was to build up such support in the country that the Liberals would be obliged to heed it when they returned to power. Stansfeld's first step was to issue an appeal to ministers of religion.

Far more, he urged, was at stake than the operation of the Acts. The fundamental question was whether prostitution should be recognised and regulated by the Government as a social institution.[25] This was followed by an appeal to the medical profession, whose attitude was crucial because many leaders of public opinion preferred to rely on the expert advice of the doctors rather than to investigate such an unpleasant question themselves. A National Medical Association for repeal was therefore formed with its own, heavily subsidised, journal.

Those who supported the Acts often quoted favourably the experience of other European countries. On the Continent state regulation of prostitution was generally supported by the medical profession, whose members influenced British delegates at international medical congresses. A British, Continental and General Federation for the Abolition of Government Regulation of Prostitution was therefore formed in 1875 with Stansfeld as President and Josephine Butler and James Stuart as secretaries.

Stansfeld himself made speeches all over the country. He often admitted that 'the respectable, religious world of this country had been to blame for the way it had looked askance from human suffering when it had been caused by this particular vice and had passed by on the other side'. There were even people inside and outside the medical profession who stated that syphilis was a blessing, inflicted by the Almighty to act as a restraint upon evil passions, and that if it could be exterminated fornication would run rampant. Stansfeld on the contrary urged that general hospitals, many of which refused to do so, should treat venereal diseases; entry to special hospitals carried a stigma which made sufferers reluctant to enter them.[26]

The first debate on the Acts in Parliament had taken place in 1870 when the Quaker William Fowler moved for leave to introduce a Repeal Bill. He withdrew his motion when the Government agreed to set up a Royal Commission on the subject. The Commission in 1871 recommended that periodical medical examination should be discontinued but that the Acts should be extended to London and other areas. In light of this report on the one hand and the impact of the repealers in by-elections on the other, the Home Secretary introduced a Bill under which the Acts would be repealed and the age of consent raised from 12 to 14. Penalties would be provided on keepers of bawdy houses. Prostitutes could be imprisoned for soliciting and detained in hospital if found to be affected by contagious disease.

The repealers' friends in Parliament supported the Bill as did at first the National Association. Mrs Butler, however, persuaded the latter to reverse its position on the grounds that the system of compulsory examination of women was shut out at one door only to be admitted by another. As the Bill was also strongly opposed on the other side by protagonists of the existing Acts, it was withdrawn by the Government. The reformers were over optimistic. Twelve years were to pass before concessions were again offered. Some of them came to regret that half a loaf had not been considered better than no bread. Parliament became bored with the question. The Liberal Harcourt Johnstone doggedly introduced a Repeal Bill each year but in 1875 it was defeated by a larger majority than it had been in 1873. When Stansfeld came to give his support the cause appeared to be faltering. He made his first major Parliamentary speech on the subject when Harcourt Johnstone's Bill of 1875 received its first reading. He started by repudiating the doctrine that diseases consequent on vice are visitations of providence and that governments have no right to assuage or cure them. A function of Christian charity, he urged, is to heal diseases. He approved of the services of hospitals and reclaiming agencies but opposed the registration and periodic examination of women 'in order that they may be returned sound on the market'. He was, he said, familiar from the inside on how governments can manipulate statistics. In fact the Acts had not reduced venereal disease but had given a false security. Since the hygienic success is disproved, he concluded, the Acts should be repealed in response to the moral instincts of a large part of the population which has raised an agitation which will never rest until it has attained its object.[27]

When the Liberals returned to office in 1880 Stansfeld was not included in the Cabinet. He was freer and of more use to the cause than if he had been muzzled by ministerial responsibility. When he took over the lead of the repealers in the Commons as Harcourt Johnstone went to the Lords, he insisted to the Association that this must be with full understanding that he would be left free to deal with proposals according to his own judgement. His most valuable contribution of all was as a member of the Committee of Enquiry into the Acts which was established by the Conservative Government in 1879, and which continued its work after the Liberals returned to office in 1880. The struggle in the Committee was largely between Stansfeld and the Government officials who repeatedly defended the operation of the Acts with statistics which he demolished. He was particularly effective in the examination of medical witnesses, from which it emerged that the Acts had not lessened the incidence of venereal disease in the Forces.

Ten members of the Committee signed its report; six members signed a minority report written by Stansfeld. The majority report was brief and complacent. The Acts, it said, had worked admirably. Indeed it indicated that it would have recommended that the system should be extended throughout the country if there had not been such strong opposition on moral grounds from ignorant people, particularly expressed through Nonconformist bodies. Stansfeld's report was argued in much more detail and with great effect. As a barrister he was experienced in dealing with the evidence of witnesses and knew how that of the police was obtained. As an expert in local government he could detect the fudging of reports. As a former minister he knew how statistics could be used and misused. Most of his report was devoted to the medical evidence. He showed how syphilis was increasing more in the stations protected by the Acts than outside them. Fear of being locked up in hospital had caused many prostitutes to leave the protected areas. Those who remained were the older ones who now each had more contacts. The medical evidence showed that after prostitutes had recovered from syphilis they often acquired an immunity but could still transmit the disease. Soldiers who frequented them in the belief that they were safe were in fact being infected by cross transmission. New recruits were particularly at risk, carelessly using a system which they regarded as provided by the Government. He pointed out that in cities whose hospitals admitted VD cases on a voluntary basis the incidence of the disease was less than in the protected areas. It was only after he had disposed of the medical case for the Acts that he asserted that the religious, moral and constitutional objections were valid in principle and confirmed by the practical results of the operation.[28]

The majority had supposed that few readers would trouble to look through the thousand pages of evidence. What Stansfeld did was to marshal that evidence with devastating efficacy. This was to be one of the rare occasions when Parliament preferred the minority to the majority report of one of its committees. The minority position was fortified by a unanimous vote in its favour by the General Committee of the National Liberal Federation. Stansfeld now moved in Parliament in 1883 that 'This House disapproves of the compulsory examination of contagious diseases.' He faced powerful opposition from three key ministers, the Secretary for War, Lord Hartington; the Secretary for the Navy, Lord Northbrook; and the Home Secretary, Sir William Harcourt. On the other hand the National Liberal Federation vote showed that in the country the party was behind him. Caught between these pressures the Prime Minister, Gladstone, allowed his ministers a free vote.

Stansfeld, in introducing his motion on 20 April 1883, tore to pieces the reasoning and the absurd claims of the majority report. He refuted the Government's statistics with those of the London Rescue Society and other independent bodies. The Acts, he declared, had led to an increase in clandestine prostitution. They had also caused a great increase in the transmission of the diseases 'through the creation of a pariah class who become the media, without themselves being infected, of conveying contagion from man to man'. The Acts had hardened women and made them more difficult to reclaim. He pleaded for free entry and exit for prostitutes to hospitals. Referring to his own part in the campaign he said, 'I have been obliged to speak largely and mainly on hygiene, but I revolt against the task. I have had the weight of this question upon me for some 10 years. I loath its details ... but never will I abandon a duty which I have once undertaken to fulfil, nor will I rest until I have proved the hygienic failure and imposture of the Acts.'[29] He was at this time under great strain imposed by the mental illness of his wife. This lasted until her merciful death in 1885.

In the debate the Conservative Cavendish Bentinck sneered at Stansfeld for his inconsistency in having been a minister in the Government which passed the Acts and having at the time preferred his peace to his principles. Samuel Whitbread rose to the defence of Stansfeld who he said had sacrificed peace, time, money and every other ambition in order to devote himself to the question.[30] Throughout the country Mrs Butler organised continuous prayer by women, led by clergy of all denominations, through the night whilst the debate continued. Cardinal Manning was in the lobby of the House, urging the Irish members to support the resolution. Stansfeld's resolution was passed by 182 votes to 110. Gladstone voted in its favour. The Government, recognising that in light of this overwhelming majority the House would refuse the estimates which provided for compulsory examination, suspended the operation of the Acts and withdrew the plain-clothes police.

The situation remained precarious. It was essential for the opponents of the Acts to follow up and achieve their complete repeal; otherwise if the Conservatives came into office they were likely to remove the suspension. No Parliamentary time was found before the Liberals went out of office in 1885. When they returned to power in 1886, as it turned out briefly, Home Rule for Ireland was the burning issue by which the party was torn apart. The repealers now had a piece of good fortune. When Hartington left the Government over Irish Home Rule his successor as Secretary for War, Sir Henry Campbell Bannerman, advised his Cabinet colleagues to vote for complete repeal. As long as the Acts remained on the statute book, he pointed out, the local authorities would not take responsibility for provision of medical treatment.[31] With the Secretary for War expressing this view, the Government supported a motion for repeal by Stansfeld in March 1886. 'Seventeen weary years had passed,' he pointed out, 'in which many hundreds of persons, both men and women, had spent their time, some their lives, and some had broken their hearts in the endeavour to get these acts repealed.' Once more he pounded the House with statistics and concluded that Parliament must finally make its decision between the principle of freedom or compulsion. Campbell Bannerman replied in an altogether lower key. He thanked Stansfeld for his clearness and temperate tone. Compulsory examination had been abolished and it was impossible that Parliament would revive it. As long as the Acts remained, he said, women would not resort to hospitals nor would local authorities exert themselves in support of hospitals. He therefore supported the motion.[32] With pressure from Stansfeld behind the scenes, and despite the turmoil over Ireland, the Bill passed through all its stages and received royal assent before the Liberals went out of office for six years.

Immediately after Stansfeld's Bill had been passed in the Commons, he returned briefly to the Cabinet as President of the Local Government Board, replacing Chamberlain who had resigned over Gladstone's Irish policy. Seeing the Irish as an oppressed nationality, he championed Home Rule in the House vigorously both before Gladstone's Bill was lost and afterwards in opposition. His last intervention in Parliament was to support a proposal that all women who would have been on the Parliamentary or municipal register if they were men should be enfranchised.

His commitment to equality for women had been expressed in many ways throughout his career. He had supported bills which proposed to extend the franchise to women and protested vigorously when Gladstone brought the party to heel in 1884 after a majority had voted in favour of Parliamentary votes for women.[33] He took a deep and effective interest in the medical education of women. When the Royal Free Hospital was opened to women students in 1877, Elizabeth Garrett Anderson wrote to him, 'we all owe more to you than anyone'. He was never petty. Dr Garrett Anderson had been one of his opponents over the Contagious Diseases Acts. He foresaw that at first other women doctors might follow her lead in this. But he said, 'it is their right and it [women doctors] must *ultimately* tell for us'.[34] He left Parliament on its dissolution in 1895 and was knighted. He died in 1898. Like Wilberforce and Shaftesbury, Stansfeld was more famous for what he achieved in influencing governments from the outside rather than from within them. But whilst they had no desire for office Stansfeld enjoyed being a minister and carrying reforms through Parliament. He was keenly disappointed not to be offered Cabinet posts in Gladstone's administrations of 1880 and 1892. He thus made a real sacrifice in becoming leader, with Josephine Butler, of a disreputable cause. As his biographers put it, 'there is nothing that men and women resent more than conduct which forces on their notice aspects which they wish to ignore.... Stansfeld came into the drawing room with the smells of the sewers hanging about his clothes.'[35]

He was remembered by his contemporaries as someone who could not forgive subterfuge; who held his opinions with religious fervour and for whom the strain and stress of a subject grossly distasteful to him undermined his happiness and peace. These qualities of rigidity and sensitivity were perhaps a defect in an aspiring politician but they were invaluable in a crusade. As W. T. Stead put it, Stansfeld had caught from Mazzini that divine thirst for self-sacrifice which enables men to give up all and follow the supreme call of duty and pity.[36]

THE JOURNALIST – W. T. STEAD (1849–1912)

In 1885 Josephine Butler and Benjamin Scott, the Chamberlain of the City of London, visited the editor of the *Pall Mall Gazette*, W. T. Stead, to seek his help. William Thomas Stead was then at the height of his fame and influence. The son of a Congregationalist minister, he had only attended school for two years and was virtually self-educated. After briefly editing the *Northern Echo* in Darlington he had come to work for the *Pall Mall Gazette* under John Morley, whom he succeeded as editor in 1883. Stead was known as the inventor of the 'New Journalism'. Lord Milner, who had been his assistant on the paper as a young man, wrote that no newspaper in any country had ever exercised so much influence as the *Pall Mall* in Stead's time.[37] Authoritarian, didactic, fervent and imaginative, he dictated his leading articles at white heat. His sudden inspirations, he maintained, came from God, whom he described as the Senior Partner. As editor of the *Pall Mall Gazette*, and later of the *Review of Reviews*, he was quite unpredictable to his friends. There was nothing however in Stead's career either before or after, of which he was so proud as the part which he was to play in the success of Mrs Butler's second great crusade.

Josephine Butler, in her work as Secretary of the International Federation for the Abolition of Government Regulation of Prostitution, had become aware of the presence of very young British girls in brothels in Belgium, to which they were able to be exported because the age of consent in Britain was only 12 years (raised to 13 in 1875) compared with 18 years on the Continent. Her concern had extended to the sufferings of young girls in brothels in Britain also. Failing to attract the attention of the Government to the need to raise the age of consent, she had threatened Lord Granville, who was Foreign Secretary and leader of the Liberals in the House of Lords, to camp on his doorstep until he would see her. When he received her he was impressed with her account, but felt that more study was required before the Government could take action. He therefore caused his friend Lord Dalhousie, who was not a member of the Government, to move for a Select Committee of the House of Lords to enquire into the law relating to the protection of young girls. The evidence heard by the Committee, and which was attached to its report published in 1882, was as sensational as the accounts which were to appear in the *Pall Mall Gazette* three years later, but like most Government blue books it was only read by a small circle. The report showed that over the past eight years British girls had increasingly been induced by agents in London to enter Belgian brothels from which escape was impossible. The Committee was 'unable adequately to express the sense of the magnitude of the evils brought to light and the necessity of taking urgent measures to cope with it'. With shame the report admitted that 'in other countries female chastity is more or less protected by law up to the age of 21. No such protection is given to English girls above the age of 13.' It recommended that the age of consent be raised to 16 and that it should be made a misdemeanour to endeavour to procure women to go overseas for the purpose of prostitution or to receive girls under 16 for purposes of sexual intercourse.[38]

In 1883 the House of Lords passed a Criminal Amendment Bill along these lines after a debate in which the aged Lord Shaftesbury, who had been a member of the Committee and had spent most of his life fighting iniquities, told the House, 'nothing more cruel, appalling or detestable could be found in the history of crime all over the world'.[39] When the Bill came to the Commons in 1883 and again in 1884, however, no time was found to consider it. When at last in May 1885 it was debated, the Home Secretary, Sir William Harcourt, supported it, saying that he had received communications about it from all parts of the country and all classes. He was warmly backed by James Stuart and Sir Baldwin Leighton. The latter indicated that the working classes were getting exceedingly impatient at the manner in which all these social questions were put on one side by Parliament. But there was concerted opposition. One

Member objected 'to making criminals out of the passions of mankind'. As for the proposed change of the age of consent, another Member argued that some girls were 'more precocious at 12 than others were at 14 and 15'. Finally the most active opponent of Mrs Butler's causes, Cavendish Bentinck, deliberately talked out the Bill. His reputation and apparent motives were such that as he sat down cries of 'fornicator' were heard.[40]

Six years had now passed since Josephine Butler had taken up the case of young girls tricked into prostitution at home or exported abroad – years during which she was haunted by the continuous evidence which came to her of their sufferings. Her efforts to put an end to these had been blocked by both Parliament and the law. To make matters worse, the Conservatives, who were even less likely than the Liberals to spare time for the Bill, had recently gained office. She came to Stead as the only person in England whose voice might break through the negligence and opposition. Stead read the dossier which she left with him incredulously. He consulted a former Head of the Criminal Investigation Department of Scotland Yard, Sir Howard Vincent, who confirmed the facts. A girl over 13, he explained, was in law regarded as consenting to her own violation once she was inveigled into a house of ill fame. A girl under that age could seldom give evidence to a judge and jury because she was not deemed to understand the meaning of an oath. Now totally convinced, Stead set up what he grandiloquently called a 'Secret Commission' to investigate the question, consisting of himself, his staff and Mrs Butler. Their findings were published in five successive issues of the *Pall Mall Gazette* early in July 1885 as 'The Maiden Tribute of the Modern Babylon'. The articles described in vivid detail how children were lured into, and kept in, prostitution in London and abroad. Their declared aim was so to arouse public opinion that the House of Commons would be obliged to pass the Criminal Law Amendment Bill which it had repeatedly shelved. 'We do not believe that members on the eve of a General Election will refuse to consider the Bill protecting the daughters of the poor, which even the House of Lords has, in three consecutive years, declared to be imperatively necessary.' Pressure was particularly concentrated on that part of the Bill which would raise the age of consent.

The evidence of the 'Secret Commission', drawn from interviews with brothel keepers, procurers and prostitutes, revealed an almost insatiable demand among middle-aged and elderly men of the upper classes for young virgins. It also described how this was met. One keeper used to go into the country disguised as a parson. He would take back a girl to London for an evening at the theatre and ensure that she missed the last train home. She was then left at a 'safe' place for the night, which was in fact a brothel. Once there, no escape was possible. Often country girls would be brought up to town understanding that they were to be domestic servants. Even simpler, in London itself young girls could be bought for between £8 and £10 from their mother. Women decoys were particularly effective. They would pick up girls on their way home from school. Well-dressed ladies would approach nursemaids pushing prams in the London parks and strike up an acquaintance by admiring the babies. Irish girls would be met from the boats which brought them to England by women dressed as Roman Catholic Sisters of Mercy. Girls arriving in London from the North were welcomed at railway stations by bogus Protestant Deaconesses. Charwomen with easy access to private houses found ways to entice children out who never returned. The accounts gave lurid descriptions of drugging and of padded cells where screams could not be heard. Throughout the articles ran the theme of class. 'All these pimps and procuresses and brothel keepers,' they said, 'are comparatively innocent. The supreme criminal is the wealthy and dissolute man.' The girls of under 16 were stated to be patronised almost exclusively by old gentlemen, including princes of the blood, Cabinet ministers and judges. Stead indicated that the names of such men in high places could be supplied if he were called on to testify.

The reaction to the articles was extreme. The *Pall Mall*'s news boys were hauled before the Lord Mayor. Cavendish Bentinck in the House of Commons demanded the prosecution of Stead for obscenity. W. H. Smith banned the *Pall Mall* from its station bookstalls. On the other hand many letters came in from those, including peers, MPs, bishops and clergy, who supported the publication of the articles, though some deplored the explicit details. Ruskin wrote from Venice, 'you have done a fireman's duty in a fire of Hell'. The quality London press, apparently on grounds of obscenity but in fact alarmed at the potential consequences of the revelations of the conduct of members of the upper classes, almost unanimously condemned the *Pall Mall*. Typically *The Times* declared 'there will scarcely be a boy or girl who will not be tainted by the disgusting publicity with which they have been so plentifully supplied'. The provincial press, on the other hand, largely edited by Nonconformists who had an innate belief in the wickedness of the Metropolis, supported Stead. Indignant readers cancelled subscriptions but the demand for copies was such that the *Gazette* ran out of newsprint.[41]

The Conservative Government was in a minority in the Commons pending a General Election, which could not be held until the electoral registers were revised in order to incorporate the voters who had recently been enfranchised. The articles seemed certain to turn Nonconformists, as well as working-class electors, many of whom were exercising their votes for the first time, against the Government. In addition, Cardinal Manning, a strong supporter of Stead, exercised considerable influence over the Irish Members. Sir Richard Cross, the Home Secretary, rapidly reintroduced the Criminal Law Amendment Act, frankly admitting that he was doing so as a consequence of the *Pall Mall* articles. Neither party could risk being branded as opposing the Bill in the forthcoming elections, nor face Stead's threat to reveal the names of those in high places who violated young girls unless the Bill were passed. It was rushed through its readings and received Royal Assent only two months after the publication of the *Pall Mall* articles and just before the dissolution for the General Election.[42]

The indignation of the establishment against Stead was particularly roused by his portrayal of the upper classes as vicious exploiters of the daughters of the poor. In his crusading enthusiasm he had played into the hands of his enemies and thus brought disaster not so much on himself – loving publicity as he did – as on his helpers. As the culmination of his revelations he had wished to provide incontrovertible proof that a child could be bought as a slave for the purpose of prostitution and exported abroad. This seemed the only way to refute the argument that children entered into prostitution voluntarily. In order to effect a purchase, a convincing intermediary was required. She was found in Rebecca Jarrett, a former procuress, reformed by Mrs Butler and now living in her Home of Rest. Rebecca was reluctantly persuaded by Mrs Butler to purchase a thirteen-year-old girl, Eliza Armstrong, for Stead from her sodden mother for £5. She was taken into the care of the Salvation Army and Stead, hideously rouged as a debauchee, rushed into her room. Then, Eliza, after medical examination to prove that she was still a virgin, was escorted by a plain-clothes Salvation Army officer to the Army's hostel in Paris. The demonstration of how the trade was conducted was complete.

The infamous Cavendish Bentinck now persuaded the mother, no doubt for a further financial consideration, to report that the child had been abducted without the consent of her husband. Much later it was revealed that 'Mrs' Armstrong was not married to Armstrong (perhaps prudently, as his only known first name was 'Basher') and indeed had never met him when the child was born. Stead, Rebecca

Jarrett, Bramwell Booth, Chief of Staff of the Salvation Army, and others who had assisted were, however, now prosecuted for abduction.[43]

The case attracted enormous attention. The Attorney General prosecuted for the Crown. Stead conducted his own defence on the hope of making sensational speeches in the guise of evidence. Most of these were cut short by the judge, but published later.[44] It was surprising that Counsel for the other defendants failed to uncover the true status of Basher Armstrong – which would have completely demolished the Prosecution case. The judge ruled that the consent of the mother was nothing and that of the father everything, but this would have been irrelevant if it was shown that there had been no marriage and that Basher could not have been the father. The jury found Bramwell Booth not guilty but Rebecca Jarrett, Stead and the others involved guilty. They added a rider recommending Stead to mercy and expressing high appreciation of the services he had rendered by securing the passage of a much needed law for the protection of young girls. There was no doubt however about the judge's sympathies. Stead was sentenced to three months in jail and the unfortunate Rebecca Jarrett to two years. Two other accomplices were also sent to jail where one of them, a Madame Mourez, died. The judge described Mrs Butler, who appeared as a witness, as 'an amiable but foolish lady'.

Stead recollected that he had never been in better spirits in his life than when he was in jail and that the only person who was unkind to him was the Prison Chaplain. After a few nights in a cell, the Home Secretary had him removed to a well furnished room in Holloway Prison where prisoners were allowed to work at their trades. Under this regulation he continued to edit the *Pall Mall Gazette*, also finding time to write a book about Mrs Butler. He was visited by MPs and other sympathisers from all over the country, and letters, telegrams and flowers came from those who were delighted by the passing of the Criminal Amendment Act. Mrs Millicent Fawcett even sent him the favourite dressing gown of her late husband Professor Henry Fawcett: 'Do not be cast down,' she wrote, 'a great movement, – as great as Wycliffe's or Luther's – has been set going, and by you.'[45]

As for the aftermath of the 'Maiden Tribute', Mrs Butler was deeply unhappy at the fate of Rebecca Jarrett, for which she felt directly responsible. She therefore wrote a life of Rebecca which did much to re-establish her reputation, which had been torn to pieces by the Attorney General as he revealed her unsavoury past. Rebecca on her release became a cheerful member of the Salvation Army. Eliza married happily and had six children, always remaining in correspondence with Stead and Mrs Butler.

The proprietor of the *Pall Mall Gazette* was somewhat shaken by the style of the Maiden Tribute campaign and its consequences. Soon afterwards Stead resigned as editor to start the *Review of Reviews*. After many more years of exciting journalism and of victories and defeats in his campaigns, he was drowned crossing the Atlantic when the *Titanic* hit its iceberg. One of his sensational articles in the *Pall Mall* twenty-six years earlier had failed to be effective. It was an imaginary account of the horrors of the sinking of a modern liner, written to draw attention to what would take place if they continued to sail with insufficient life boats. As the *Titanic*'s scandalously few boats and rafts, which were only sufficient to take off the women and children, departed, Stead was reported by a survivor as 'preserving the most beautiful composure and in a prayerful attitude of profound meditation'.[46] Perhaps he was being consoled not only by his Senior Partner but by the knowledge that, for the last time, he had been proved right in one of his campaigns.

Of Josephine Butler's three most influential male allies, one had forfeited the prospect of ecclesiastical promotion, one had sacrificed a political career and another landed in jail. None of them regretted their involvement.

4 Emancipation through Birth Control

We utter an emphatic warning against the use of unnatural means for the avoidance of conception... and against teaching which under the name of science and religion encourages married people in the deliberate cultivation of sexual union as an end in itself.

Encyclical of Lambeth Conference of Anglican Bishops (1920)

The love envisaged by the Lambeth Conference is an invertebrate, joyless thing – not worth the having.

Lord Dawson of Penn at the Annual Church Congress, Birmingham (1921)

Whilst some of the middle-class women who led and supported the feminist movement in the nineteenth century remained unmarried, most of those from the middle classes hoped to be able to combine fulfilling careers with family life. It was hard to do so at a time when births were unchecked and married women often lived in a state of almost permanent pregnancy through what could have been the most important years of a career, and not infrequently died of the consequences. Nothing did more for the liberation of women in the nineteenth and twentieth centuries than the spread of contraception. The principal obstacles were ignorance, the opposition of the Church and the hostility of the medical profession. The men who took the lead in overcoming these at the cost of unpopularity and persecution deserve to be remembered as notable heroes of the early feminist movement.

How contraception came about varied considerably in different countries. In France women in the aristocracy and bourgeoisie developed home-made devices in the eighteenth century. These came to be used in the early nineteenth century in the peasant and farming class also, mainly because under the Code Napoléon each child had to be left a share in the estate; large families would thus lead to uneconomic holdings. Knowledge spread by word of mouth; there was no propaganda. The French birth rate fell considerably before that in other European countries, though eventually the prospective loss of military manpower caused a right-wing and Catholic coalition government in 1920 to pass a law prohibiting the dissemination of contraceptive information. By then, however, the practice was too well established to be checked and the opposition to it of the Roman Catholic Church appears to have had little effect.[1]

FREETHINKERS AND PIONEERS

By contrast, in Britain and America birth control started to be practised as a result of propaganda campaigns, led until the end of the nineteenth century mainly by men, who were sometimes fined or imprisoned and whose work was first ignored and later condemned by the Church and the medical profession. Its earliest British proponents were freethinkers, strongly opposed to the Establishment; all of them were feminists, and birth control was only one of the many causes which they championed.

The first of the propagandists was Francis Place (1791–1854), the Radical tailor. He had 15 children, 5 of whom died in infancy; in his early married life the family lived on the verge of starvation. Eventually he prospered, making elegant coats for arrogant aristocrats, and was then able to give up his trade and devote himself to reform and working-class politics. His experience of poverty coloured all his writings. His interest in contraception was aroused while he was writing a book to confute Malthus. The latter had maintained that population would outstrip food resources unless the increase could be controlled by late marriage and abstinence; Place considered that the welfare of labourers could only improve if their numbers increased more slowly, but that Malthus's remedies were unrealistic. He now wrote and had distributed three contraceptive handbills addressed respectively 'to the married of both sexes', 'to the married of both sexes of the working people' and 'to the married of both sexes of genteel life'. These, which came to be known as the 'Diabolical Handbills', explained that large families led to poverty, and they gave practical information on the means then used in France to prevent conception. It was during the distribution of the handbills, as has been seen, that the young J. S. Mill was arrested. The reputation which Place gained as a result of this campaign caused him to be excluded from any leading position in the Chartist Movement, which he had done much to bring into being.[2]

Place also stimulated the writer and publisher Richard Carlile to an interest in the subject. Carlile (1790–1843) had started life as apprentice to a tinman. Self-educated, he later became a writer and publisher.

He spent many years in prison and did more than any other Englishman of his time for the cause of freedom of the press. When Place wrote to him in Dorchester jail he was serving a sentence for blasphemy but continued to edit his paper *The Republican*. His first reaction was that contraception was obscene. When Place convinced him that it was not, he published in *The Republican* a long article, which subsequently became a pamphlet, with the title '*What is Love*?' He wrote this, he said, in response to a flood of letters asking for information on the subject. The pamphlet, published in 1826 as *Every Woman's Book*, explained methods of contraception in more detail than Place had done. It also argued the case for it more broadly. The practice, he suggested, would prevent unhealthy women from endangering their lives and would become 'the very bulwark of love and wisdom, of beauty, health and happiness'. Carlile's books were sold from a clockwork rotating machine in order that police agents could not identify the vendor; 10,000 copies of *Every Woman's Book* were sold in the next four years.[3] One of these reached Robert Dale Owen, son of the more famous social reformer Robert Owen, in New York. After consultation with Place and Carlile, whom he visited in England, Owen followed up their work in 1831 with a book *Moral Physiology, or a Brief and Plain Treatise on the Population*, described by his biographer as 'the most temperate, refined and readable of the 19th century tracts on birth control'. Whilst it contained nothing new about contraceptive practice it was important for its social and eugenic arguments in favour of family limitation. It had a much larger sale in Britain and America than Carlile's book. Its authorship did Owen no harm, for a few years later he was elected to the Indiana State Legislature and then to Congress. In his subsequent political career however, he found it prudent to champion other causes.[4]

A copy of Owen's book in turn came into the hands of Dr Charles Knowlton who became the first person to go to prison for advocating birth control. Knowlton (1800–50) was a backwoods physician in Massachusetts, who grew up on his father's farm. His practice convinced him that too frequent childbirth impaired the health of wives as well as imposing excessive financial burdens on young couples. Starting by writing a manuscript leaflet giving contraceptive advice to his patients, in 1832 he felt that this needed wider circulation and published it anonymously in New York as *Fruits of Philosophy, or the Private Companion of Young Married People by a Physician*. The book set out to answer the contemporary argument that contraception was opposed to nature. It maintained that the sexual drive was a normal human appetite but that its uncontrolled exercise led to frequent births which caused the ill health and early death of mothers. Conception control, he said, would lessen prostitution by making early marriage possible, improve the health of both mothers and children, and make abortion and infanticide less common.

This led to a brief – and for its time fairly accurate – description of the reproductive process and the means for checking it. Nowhere however were the old Puritan ordinances still enforced more stringently than in Massachusetts. Knowlton was prosecuted three times as writer of the book. On the first occasion he was fined, on the second imprisoned for three months and on the third the jury disagreed. He was not a crusader. His remaining life appears to have been uneventful; accepted by the community, his writing was now confined to unprovocative topics in professional medical journals. It was only long after his death that he became famous.[5]

In England the birth control cause was now carried forward by Dr George Drysdale (1825–1904) in a massive book, *Physical, Sexual and Natural Religion*, published in 1855, which attacked the Church as the enemy of science and devoted only five of its pages to practical advice on how to avoid conception. Drysdale argued that any preventive means, to be satisfactory, must be used by the woman. He saw sex as a bodily need and believed that the sexual organs required regular use to avoid atrophy. He combined science, insight, hedonism and common sense in a way that no one had previously attempted.[6]

It was Charles Bradlaugh and Annie Besant, however, who by republication of Knowlton's book and their consequent trial in 1877 gave such publicity to the subject as to bring about a significant fall in the British birth rate. Bradlaugh (1833–91) was the son of a solicitor's clerk. As a youthful Sunday School teacher he lost his faith in Christianity. Though he had little formal education he acquired a knowledge of law which was to be a useful asset in his stormy career. The most prominent freethinker of his time, he gained great notoriety by repeatedly refusing to take the required oath when elected to Parliament.

Bradlaugh was a passionate believer in free speech. In 1876 a Bristol bookseller was sentenced to two years' hard labour for republishing Knowlton's *Fruits of Philosophy* and adding 'obscene' illustrations to it. Thereupon Bradlaugh and Annie Besant formed a Freethought Press and again republished the book with medical notes by Dr George Drysdale. Prosecution was virtually invited by sending copies to the Lord Mayor of London and the police. In the previous four decades only 700 copies of it had been sold. Now in the first twenty minutes 500 copies were bought and sales rose to 125,000 in the three months between the arrest of Bradlaugh and Mrs Besant and their trial. The authors were prosecuted by the Crown, probably at the instigation of the Society for Suppression of Vice. Bradlaugh printed a public statement saying 'I hold the book to be defensible and I deny the right of anyone to interfere with full and free discussion affecting the happiness of the nation.' Drysdale on behalf of the defendants described how working-class women suckled infants for as long as two years in order to prevent conception; the consequence was that the child was deprived of its proper food and, if the woman conceived again, the unborn child suffered from want of nourishment. The Solicitor General described the reprint as a 'dirty and filthy book which no decently educated English husband would allow his wife to have'. The jury found that it was calculated to deprave public morals but exonerated the defendants from any corrupt motives in publishing it. Bradlaugh and Mrs Besant refused to undertake to cease to sell the book and each was sentenced to six months' imprisonment.[7] Their appeal was successful on a technical point. After considerable hesitation the government decided not to prosecute further. At about the same time however, the Society for the Suppression of Vice caused the police to raid the premises of the publisher Edward Truelove and confiscate 219 copies of Robert Dale Owen's book. Truelove was sent to prison for four months.[8]

Bradlaugh in his radical Parliamentary constituency of Northampton was to be unharmed by his association with birth control. For Viscount Amberley, the son of the former Prime Minister Earl Russell, and MP for South Devon, involvement was to be politically disastrous. Chairing a meeting of the London Dialectical Society in 1868 on 'the happiness of the community as affected by large families' he was reported as saying that 'unfortunately the influence of the clergy was opposed to the prevention of overpopulation', and also 'that he wished he could hear the proposals of medical men as to the best means of limiting numbers'. The *British Medical Journal* commented, 'We believe that our profession will repudiate with indignation and disgust such functions as these gentlemen wish to assign to it.' Its rival the *Medical Times and Gazette* deplored Lord Amberley's 'most scandalous insult' and declared that 'the day would never come when an obscene branch of trade now limited to dens of obscenity and vice would be fostered and protected and introduced into homes by the medical profession'. In a subsequent election campaign, local and even London newspapers took up the cry from the medical press, and the Roman Catholic Archbishop of Liverpool attacked Amberley in a sermon. Tory opponents called him the 'Vice Count' and he lost his seat. His son, Bertrand Russell, was influenced by his father's example in his own later support for birth control.[9]

THE MEDICAL PROFESSION

The attitude of the medical profession towards contraception in the nineteenth century was strongly influenced by its preoccupation with respectability. Slowly its members had climbed from being barber surgeons, then apothecaries who ate with the servants, to become recognised as gentlemen, wearing frock coats and carrying a string of initials after their names. Contraception was associated in their own as well as in the public mind with quacks who advertised and peddled their dubious wares in the margins of legality.

The doctors were taunted by the radical advocates of birth control for refusing to interest themselves in the problems of the poor and for having knowledge which they kept to themselves. As G. J. Holyoake pleaded, 'We want a physician of authority to treat the medical view of the question with the same power and boldness with which Mill treats its bearing on political economy.'[10] He pleaded in vain. There were a very few mavericks in the profession such as those of the Drysdale family who recommended the use of contraceptives, but these were boycotted by their colleagues. A worse fate befell Dr H. A. Allbutt, Consultant to the Institute for Skin Disease in Leeds. He published in 1886 a sixpenny pamphlet *The Wives' Handbook*, whose purpose was 'to save the life and preserve the health of hundreds of women and to teach the young wife how to order her life during the most important period of her life and to remove from her the popular ignorance in which she may have been raised'. The pamphlet contained a short chapter on contraception and stated that the safe period technique failed in about 5 cases in a hundred. Advertisements for contraceptives were printed at the end. The General Medical Council struck Allbutt from its register for 'infamous conduct in a professional respect' in having published the book at so low a price as to bring it within the range of youth of both sexes. The inclusion of the advertisements no doubt compounded the offence. The *Pall Mall Gazette* described the action of the Council as 'one of the most glaring illustrations of professional prejudice and human folly'. As in the Bradlaugh–Besant case the publicity led to huge sales. Half a million copies had been sold by 1926.[11]

Whilst in general the policy of the medical journals and leaders of the profession was to say as little as possible on contraception, the Amberley affair moved the *Lancet* in 1869 to declare that the wife in a marriage in which contraception was practised 'is necessarily brought into the condition of mind of a prostitute'. In the same year Dr T. E. Beatty in an address on midwifery at the British Medical Association's annual meeting blamed 'the odious doctrines openly recommended by societies composed of the upper ranks of the people, and presided over by senators, as leading to criminal abortion'. He suggested that the Dialectical Society should change its name to 'Diabolical'.[12] Privately and in advising their patients, a number of doctors came to accept that spacing of children was desirable but only to be implemented by abstention, by prolonged nursing, or by the rhythm method, whose complexity and fallibility was little understood at the time.

At the turn of the century the attitude of the profession changed from ignoring birth control to attacking it publicly. The change was partly due to the wave of Imperialism which was at its high tide. Contraception was widely described as leading to race suicide. Dr J. W. Taylor, President of the British Gynaecological Society, in an article in the *Nineteenth Century* in 1906, reflected on recently published tables showing the simultaneous decline in the birth rate and death rate in England and Wales. He wrote that 'No modern surgeon or physician when reviewing his life work can contemplate this table with any other sentiment than that of supreme dissatisfaction and disgust. All benefits of modern medicine are swept away by the present generation.' Birth prevention, he sweepingly declared, brought grievous medical, physical, moral and social evils on the community. In his Presidential Address to the Society in 1904, he complained that 'The highest and noblest function of the married woman, the rearing of sons and daughters to the family, the nation and the Empire, is very largely handed over to the lower classes ... and to the Hebrew and the alien.'[13]

Several of the medical witnesses to the National Commission on the Declining Birthrate in 1916 favoured spacing of children: none, however, endorsed the use of artificial methods. Dr Amand Routh, Consultant Obstetric Physician to the London Hospital, stated, 'I have no doubt that prevention of maternity by artificial means always does harm. The nervous system suffers enormously if the habit continues for long.' Interestingly, in his experience the initiative in refraining from having children generally came from the wife. He invoked biblical authority for the unrestricted growth of population.[14] Dame Mary Scharlieb, the doyenne of the obstetricians, told the Commission that contraceptives should be prohibited. She too spoke of the damage to the nervous system of contraception and indignantly exclaimed that, 'The whole nation needs a bath of physiological righteousness – the whole nation is rotten from top to bottom.'[15] The clerical witnesses too were unanimously opposed to the use of contraceptives. That the attitude of both the medical and ecclesiastical professions appeared to come close to hypocrisy was revealed by another witness, the Superintendent of Statistics in the General Register Office, who in his evidence informed the Commission that the professions whose members had the lowest birth rates were, in order, doctors, teachers, Church of England clergy, and then other clergymen. Some demographic historians have suggested however that the decreasing birth rates among these classes was largely due to 'the culture of abstinence'.[16]

LORD DAWSON OF PENN VERSUS THE CHURCH OF ENGLAND

The freethinkers were natural foes of the Church of England. Carlile, for example, wrote that 'The only enemies my book should find are royal families, aristocracy and priests', and kept in his front window an effigy of a bishop arm in arm with a devil.[17] Throughout the nineteenth century the Church's attitude to birth control was generally to ignore the subject in public and condemn it in private. As hardly any biblical texts appeared relevant, except the general injunction to be fruitful and multiply, it tended to rely on St Augustine who wrote that sexual intercourse with a lawful wife is unlawful and shameful if the offspring of children is prevented, and on St Thomas Aquinas' elaboration of this. The Roman Catholic position was always one of complete hostility to all artificial means of birth control. This was eventually formulated in the Encyclical of Pope Pius IX, Casti Conubii, in 1931, which condemned contraception as 'shameful and intrinsically immoral, an unspeakable crime'. Several subsequent Papal Encyclicals endorsed this. The firm position of Rome affected the attitude of the Anglo-Catholic wing of the Anglican Church, which yearned for reunion with Rome and opposed any obstacles to this. Even F. W. Farrar, Dean of St Paul's, who was considered a progressive, wrote in the *Fortnightly Review* in 1888 that the Church of England recognised no remedies for overpopulation except abstinence.[18]

In the early twentieth century the Church of England, like the doctors, came out in open condemnation. The new Bishop of London, A. F. Winnington Ingram, in his primary episcopal visitation in 1905 told his clergy that it was 'impossible for him to describe with what dismay he viewed the diminution of the birth rate . . . and the proved cause to be its limitation by artificial means. To stem this gigantic evil,' he declared, 'I summon the forces of the Church today against a practice which must eat away the heart and drain away the life blood of our country.'[19] His tone was echoed in the Encyclical Letter of the bishops of the Anglican Communion at the Lambeth Conference of 1908 –

> A further evil of which we have to deal is of such a kind that it cannot be spoken of without repugnance. . . . In view of the figures which have been set before us we cannot doubt that there is a widespread prevalence amongst our peoples of resorting to artificial means of the avoidance of childbearing: we would appeal to the members of our own churches to exert the whole force of their Christian character in condemnation of them.

In the report of its Committee on Marriage Problems, the same conference asserted that the use of artificial means of prevention was associated with nervous enfeeblement and insanity. It warned of the danger that 'the great English speaking peoples, diminished in number and weakened in moral force, should commit the crowning infamy of race suicide and so fail to fulfil that high destiny to which in the providence of God they have been manifestly called'. The report urged that the sale and advertising of contraceptives should be prohibited and that those who publicly and professionally assisted preventive methods should be prosecuted. The conference expressed most cordial appreciation of 'the services rendered by those medical men who have borne courageous testimony against the injurious practices spoken of'.[20]

It is hard to find anyone prominent in the Church of England before 1930 who supported birth control out of sympathy for women. Dean Inge, better known for his columns in the *Evening Standard* than as a pillar of the Church, favoured birth control for eugenic reasons. Bishop Barnes of Birmingham, who was regarded by his colleagues as a wild appointment by Ramsay MacDonald and almost as a heretic, was also mildly interested for similar reasons. The young Stewart Headlam supported Bradlaugh and Annie Besant at their trial, but he was an eccentric, four times successively dismissed as a curate and finally unable to obtain a licence.

It might have been supposed that the 1914 war, in which respect for authority and tradition had been gravely shaken, would have made the Anglican bishops more cautious. Yet the Lambeth Conference of 1920 once more uttered an 'emphatic warning against the use of unnatural means for the avoidance of conception with the grave dangers, physical moral and religious, thereby incurred and the evils with which the extension threatens the race'. It also affirmed the opposition of the Anglican Church to the 'teaching which under the name of science and religion encourages married people in the deliberate cultivation of sexual union as an end in itself'.[21]

It was this last sentence which, according to his biographer, made the impact on Lord Dawson of Penn, Physician to the King, of a red rag to a bull and caused him to deliver a devastating lecture which did more than anything both to expose the obscurantist attitude of the Church to birth control and to bring about its acceptance by the medical profession. Bertrand Dawson (1864–1945) was the son of an architect who was also a lawyer. There were seven children in the family, brought up in a typically middle-class Victorian evangelical atmosphere. Bertrand went from St Paul's School to study science at University College, London. Here he became a liberal – even a radical – ardently in favour of Irish Home Rule, admiring Bradlaugh for his fight to take his seat in Parliament by affirmation instead of by oath, and questioning the dogmas of the Church of England. When he went on to study medicine at the London Hospital he developed an intense interest in Neo-Malthusianism, as birth control was then generally known. In it, says his biographer, he found 'the spirit of a social crusade, the simple desire to do good, the interest in human problems, the excitement of a deliverance from prudery and reaction, the exaltation of practical ethics as against dogma, the search for scientific truth and the desire to invest the profession of medicine with the aspect of a mission'.[22] He wrote to Dr C. S. Drysdale that the furtherance of the cause depended more on medical men than on any other class in the community.

He tabulated all the discoverable methods of contraception and made exhaustive enquiries as to their efficacy and suitability. He studied the sociological and to some extent the legal aspects. With a group of friends from University College and the London Hospital he produced a leaflet, *Few in the Family, Happiness at Home*, addressed to the poorer classes. Its message was the desirability of spacing of children in the interest of the welfare of the whole family, and the means of ensuring this. The boldness of the young men in taking up this cause can be appreciated when it is recognised that they were writing their pamphlet just two years before Dr Allbutt, with whom Dawson greatly sympathised, was struck off the medical register for publishing his book on birth control. After Dawson qualified, his advance in his profession was rapid. As well as conducting a successful practice in Harley Street, he was a consultant at the London Hospital, which had a close connection with the Royal Family. Sir Frederick Treves, Dawson's mentor, had dramatically performed one of the earliest ever appendectomies on King Edward VII just before his expected coronation, and Queen Alexandra was an active patron of the Hospital. At the gargantuan house parties of the Edwardian era a doctor was usually included in case a guest collapsed. The presentable Dawson was at one of these when his services were required for a close friend of the King, who anxiously discussed the case with him. Shortly after this Treves successfully recommended him to be appointed as one of the King's physicians. In this capacity he was knighted in 1911. In the war of 1914–18 he served as Consulting Physician to the British Army in France, with the rank of Major-General.

After the war he gained the ear of Lloyd George, who was looking for ways to redeem his election pledge of making Britain a place fit for heroes to live in, by his advocacy of a separate health ministry. He was appointed chairman of a committee set up by the new ministry, whose report was the first to draw attention to the need for a systematised health service and contained many of the ideas which led to the creation of the National Health Service 30 years later. Lloyd George, impressed by Dawson's unusual gift for organisation and of harnessing different interests to the same purpose, made him a baron in 1920, not so much for past services as because 'he wanted to have in the House of Lords some real authority on public health'. In his letter of thanks to the King's Private Secretary, Dawson recognised that 'henceforth medicine must play its part in planning and executing a health policy for the nation and to effect this doctors will need to take a share in public affairs in addition to their present duties'.[23]

Thus when Dawson was invited to give an address to the Church Congress in Birmingham on sexual relationships in 1921 he could speak with unique authority as the recognised leader of his profession, whose medical practice included many of the most eminent people in the Establishment but whose voluntary work at the London Hospital also gave him an insight into the health problems of the poor. He was, too, a member of the Church of England.

In his address he bluntly and movingly challenged the pronouncement of the Bishops at the Lambeth Conference of the previous year which condemned sexual love as such and sanctioned it only as a means to one end, procreation. 'Think of the facts of life,' he urged,

> let us recall our own love – our marriage, our honeymoon. Has not sexual union over and over again been the physical expression of our

> love, without thought or intention of procreation? Have we all been wrong? Or is it the Church that has lost that vital contact with life which accounts for the gulf between her and the people? The love envisaged by the Lambeth Conference is an invertebrate, joyless thing – not worth the having. Sex love has, apart from parenthood a purpose of its own. It is an essential part of health and happiness in marriage. If sexual union is a gift of God, it is worth learning to use it. It should be cultivated so as to bring physical satisfaction to both and not to merely one.

The last sentence was a sharp contradiction of the accepted doctrine of both the medical and clerical professions that women did not, or should not, enjoy sex.

Birth control, Dawson continued firmly, was here to stay. Absence of birth control meant a strain hostile to health and happiness. It was no more unnatural to control conception by artificial means than to give anaesthetics in childbirth. Finally, with a disavowal of the traditional refusal of responsibility by his profession, he asserted that the advantages and disadvantages of various contraceptives was a technical matter for doctors to determine.[24]

A wave of publicity followed. The *Sunday Express*, under the headline 'Dawson Must Go', urged the King to dismiss him as Royal Physician. The *Spectator* on the other hand described the speech as wise, bold and humane. The Roman Catholic Bishop of Liverpool and several Anglican bishops attacked it. The Archbishop of York, Cosmo Gordon Lang, said that it was 'very unguarded'.[25] Lang's remarks gave Dawson the opportunity to reply vigorously in an introduction to the printed version of the address. What he had said, he declared, had not been careless but outspoken. What was required was reasoned consideration, not unreasoned condemnation. There was no scriptural sanction for the Church's condemnation nor did the report of the Lambeth Conference give any valid reason why birth control was physically or morally harmful. Moving into the attack, he asserted that the attitude of the Church towards the problem of sexual relationships was part of the larger question of the separation of its beliefs from those of the younger generation. Since the war, youth had become estranged from the Church and, whilst possessed of deep religious feeling, was trying to build up a superstructure more in accordance with the progress of revelation not only in religion but in science.[26]

The Times in reviewing the printed version of the address described it as having aroused 'immense interest', and praised the passage in which

Dawson accused the Church of being out of touch with youth.[27] What gradually became clear was that Dawson, with his unique authority, had demolished the medical objections to birth control put forward not only by his professional colleagues but, even more imprudently, by the Church to justify its case. At the next Lambeth Conference in 1930 the bishops were divided. A resolution on Life and Witness of the Christian Community was the rare occasion of a vote; carried by 193 to 63 it declared that 'the sexual instinct is a holy thing, implanted by God in human nature'. Though continuing to favour abstinence, the resolution admitted that 'in those cases where there is such a clear felt moral obligation to limit and avoid parenthood, and where there is a morally sound reason for complete abstinence, other methods might be used'. To mollify the opponents, the conference strongly condemned the use of conception control from motives of selfishness, luxury or mere convenience.[28]

The opponents were not reconciled. The saintly Bishop Gore is described in his retirement by his biographer as 'reduced to indignation and almost despair, quite overcome with grief' at the acceptance of an 'unnatural, immoral and ungodly' practice.[29] The more rumbustious Bishop Furse of St Albans continued to attack contraception as 'unnatural, repellent, degrading and unsound and against God's plan'.[30] The irrepressible Dr Winnington Ingram, Bishop of London, told the House of Lords in 1934, 'I should like to see British families spreading all over the earth, multiplying abundantly' and that he would also like to make a bonfire of contraceptives ('filthy things') and dance round it.[31] But the resolution was criticised too from the other side. Bishop David of Gloucester took issue with its statement that the primary and obvious method is complete abstinence from intercourse. The Church, he said, should use her freedom to bring new knowledge to bear in new conditions. 'Rome attempts to regulate – we trust.'[32]

This indeed was the crux. The Church of England, unlike Rome, was not bound by its past. It was free to change its position on birth control just as it was free quietly to jettison the Athanasian Creed, which had caused so much unhappiness to its members in the same period. Eventually under the gentle guidance of an American bishop the Lambeth Conference in 1968, long after Dawson's death, virtually moved to his position, accepting the 'sacramental' nature of sexual intercourse, and family planning as an important factor in family life. It was utterly wrong, the conference now acknowledged, to urge that unless children were specifically desired sexual intercourse was in the nature of sin, and it expressed its firm disagreement with a recent Encyclical of the Pope which condemned contraception.[33]

Among the medical profession Dawson's address also turned the tide. A year after it was delivered a questionnaire to gynaecologists found that three-quarters of them favoured use of contraceptives. Roman Catholic physicians protested noisily but in 1933 Lord Horder, physician to the Prince of Wales and President of the newly formed Birth Control League, felt able to tell the League

> The courage with which the pioneers laboured and taught is no longer necessary: but the persistence which they displayed must still characterise the efforts of the apostles of this gospel if their convictions are to spread and their work is to bear fruit.[34]

BRITAIN AND AMERICA

The British advocates of birth control could feel by the 1930s that the opposition of the medical profession and Church had been largely overcome, and they were not restricted in their activities by the law. The parallel story in the USA was quite different. After the prosecution of Dr Knowlton in 1832 there seems to have been less overt opposition to birth control than in Britain and less distrust of the practice in the medical profession. Several medical journals organised symposia on the subject.[35] Progress was then paralysed however by pressure on Congress from the Roman Catholic Church and from the fanatical Anthony Comstock, Chief Special Agent for the Society for the Suppression of Vice. In 1873 the Comstock Law was enacted, under which anyone who sent through the Federal mail contraceptives or literature explaining or advocating contraception was made liable to five years' imprisonment and a fine of $5,000. In 1876 Dr Edward B. Foot was fined $3,000 under the Act. In all, the indefatigable Comstock was responsible for the arrest of 3,873 persons under the Act, of whom 2,911 were convicted and 333 served jail sentences.[36] The medical profession, many of whom were Roman Catholics, made little protest until about 1915 when Dr William J. Robinson in one of his books included a chapter on contraception consisting entirely of blank pages and described the censorship as worse than any which ever existed in darkest Russia. About the same time the 80-year-old Dr Abraham Jacobi, in his Presidential Address to the American Medical Association, caused something of a sensation by speaking, though in guarded language, in favour of contraception.[37]

Dr Robinson's blank pages and the flight of Margaret Sanger in 1914 to take refuge in England when prosecuted for writing about birth control caused wide uneasiness, particularly when a letter was sent by a number of eminent British intellectuals to the President of the United States, urging that her prosecution be dropped. Was the Mother of Liberty, it was asked by many Americans, to be shown up as a tyrant by the former colonial power? President Wilson may himself have been responsible for the abandonment of the prosecution of Mrs Sanger on her return.[38]

In 1929 the New York police, at the instigation of the office of the Roman Catholic Archbishop of New York, raided Mrs Sanger's clinic, herded the medical director and staff into a police wagon and took them to jail.[39] The waiting patients were made to give their name and address and the case records were carried off. Similar raids occurred in clinics as late as 1937 in Massachusetts and 1939 in Connecticut; the latter followed a resolution of censure on contraception which was read out in all Roman Catholic pulpits in the diocese. Mrs Sanger and others lobbied for years to try to persuade Congress to repeal the Comstock Acts, but few Senators and Congressmen, however personally sympathetic, dared to offend the Roman Catholic hierarchy, under whose influence some Congressmen even described the birth control movement as a communist plot. In the end it was not the legislature but the higher judiciary, whose members had security of tenure, which drew the teeth of the Comstock Law in a series of judgments in the 1930s. These first ruled that its penalties could not apply to medical practitioners and then that its scope had not been understood when it was passed.[40]

The birth control cause was never a mass movement. Its early proponents such as Place, Carlile and Bradlaugh supported birth control not only as feminists but as champions of freedom of speech and enemies of censorship. Themselves outside the Citadel, their provocative behaviour hardly helped the movement to find allies within it; Carlile called doctors 'as wicked a set of impostors as priests'. These radicals were as likely to be jailed for blasphemy or political offences as for obscenity, and sometimes they welcomed prosecution for the notoriety which it brought.

Whilst the radicals had little to lose by their involvement, perhaps the real heroes of the movement were those few members of the medical profession, such as Knowlton and Allbutt, who risked imprisonment and ostracism by their colleagues in a cause in which the health of their women patients provided the main impetus. At a later stage Lord Dawson made a major contribution in changing the attitude not only of the Church and medical profession, but eventually of Parliament, the press and the general public. His position, however, was virtually impregnable.

Curiously, unlike Mrs Butler's campaigns, and that for the suffrage, there was little need to influence Parliament in Britain, compared with the United States where the campaigners had to spend much time lobbying Congress and appealing to the courts. In Britain where it was hard – almost impossible – to find support for birth control among the clergy – the influence even of the Established Church, which repeatedly called for banning the sale and distribution of contraceptives, had no impact on legislation. Paradoxically, in America, where there was no established church, the Roman Catholic hierarchy was much more effective in blocking birth control.

Finally the supportive role of the husbands of the women who led the birth-control campaign in its final stage deserves to be remembered. Its best known leaders in the twentieth century were Marie Stopes in Britain and Margaret Sanger in America. Each had an unsuccessful first marriage and then married a wealthy man whose contribution was important at a crucial point. Margaret Sanger's husband, Jonah H. Slee, was president of the Three-in-One Oil Company and his financial support, according to her biographer, 'transformed private philanthropy into a national movement'. Further, to avoid the ban on sending contraceptives through the mails, he imported them as contraband in oil drums from his company's Canadian office.[41] Marie Stopes' second husband, Humphrey Roe, was managing director of an aviation business. He had already developed a strong interest in birth control through his experience of the Manchester slums, where he planned to open a clinic before he met Marie. Instead he financed the one which she opened in London. He became Secretary to the Society for Constructive Birth Control under Marie Stopes as President, and addressed meetings all over the country on its behalf. After he lost most of his money in the financial crash of 1929 and also his sexual drive, Marie Stopes had little further use for him. She made him take his meals in a separate room, humiliated him in front of a lover who was 33 years her junior, and eventually expelled him from their house. It is hard to think of anyone who brought happiness to so many people by their work and writings, yet whose character was so unpleasant as that of this vain and selfish woman.[42]

5 Gandhi and Liberation of Women through the Freedom Movement

> *The movement is essentially a movement for freedom of the women. The full freedom of India will be an impossibility unless your daughters stand side by side with your sons in the battle for freedom, and such an association is not possible on absolutely equal terms on the part of India's millions of daughters unless they have a definite consciousness of their own powers.*
>
> M. K. Gandhi (1927)[1]

There is no one perhaps in modern times about whose life and ideas so much has been recorded and written as that of Mahatma Gandhi (1869–1948). His collected works alone fill 90 substantial volumes. He would usually speak and write quite spontaneously, and complete consistency cannot reasonably be expected from someone whose autobiography was called *The Story of My Experiments with Truth*. He told a meeting of the Gandhi Seva Sangh (Gandhi Service Society) in 1936,

> There is no such thing as Gandhiism. I do not claim to have originated any new principle or doctrine. I have simply tried, in my own way, to apply the eternal truths to our daily life and problems.... The opinions I have formed and the conclusions I have arrived at are not by any means final. I may change them tomorrow if I find better ones.

He used to say, 'It is not what I write but what I do that matters.'[2]

To try to understand his attitude to the role of women may be a bewildering exercise unless three threads which ran right through his long public life in India are discerned. Firstly from his late thirties he was a Brahmachari, himself pledged to abstention from sex and maintaining that the only justification for sexual activity by anyone was the duty to have children. This abstinence could be used to conserve spiritual power to serve the wider family of mankind. Secondly there was no separation between his approaches to India's political, social and economic problems; the fight for *swaraj* (freedom), he said, meant not only political awakening but an all-round awakening, social, educational, moral and political, leading to a total transformation of society. Thirdly, despite his interest in and borrowing from other religions, he was a devout Hindu.

When Gandhi returned finally to India in 1914 at the age of 45, his attitude to the capacities of women had been considerably influenced by his experiences abroad. In England the shy 18-year-old student who hesitated to admit that he already had a wife and son in India had been laughed out of his embarrassment by kindly ladies who provided substantial vegetarian Sunday lunches and by their jolly daughters who took him for hearty walks afterwards. Mixing with Theosophists, he admired their leaders, Annie Besant and Madame Blavatsky. He sat on Vegetarian Society committees with young women of advanced views. Later in South Africa during the war of 1899–1902 he described the Boer women as showing spirit such as no other women had ever done. Europeans, including Jewish and Quaker women, had worked in his settlements and in his campaigns in South Africa, and he had mobilised Asian women there to cross the border illegally in a protest against laws which treated their marriages as invalid. Visiting England in 1906 he had praised the women who had demonstrated outside Parliament to demand the vote, enduring insults from the public and brutality from the police, though he lost sympathy with the Suffragettes when they turned to violence. In general his observations led him to conclude that in all countries where people lived a decent life there was no disparity between the condition of men and women.[3]

In the India to which Gandhi returned, some gradual improvements in the position of women had been achieved over the past century through the efforts of Indian reformers who were mostly inspired by British liberals and radicals and were sometimes associated with missionary educationists. As early as 1829 Lord William Bentinck as Governor General had been persuaded by Ram Mohan Roy to make *suttee* (the burning of widows) illegal. Some progress had been made in female education and in legislation regarding the age of consent. Yet several of the Indian reformers had set a bad example in yielding to social pressure by marrying off their daughters as children or even marrying children themselves. After the uprising of 1857 the government became cautious about interfering with social customs, and towards the end of the century tended to rely on the support of conservative elements who were opposed to social change. One authority has suggested that it is doubtful whether in a hundred years of social reform as much would have been achieved as Gandhi was to bring about in twenty years.[4]

For a very small proportion of Indian women advances had come almost simultaneously with those in Britain. They had been admitted to universities and had qualified as doctors in the 1880s; Cornelia Sorabji became India's first woman lawyer in the 1890s. Women became eligible to vote for the provincial legislatures on the same restricted suffrage as men from 1921, though so few had the necessary educational or property qualifications that only 1 per cent were able to do so. From its inception in 1885 a few women attended the annual meetings of the Indian National Congress. Its founder, the former Indian civil servant A. O. Hume, had seen Congress as a vehicle for social reform. With this emphasis it might have done more for women, but the Viceroy, Lord Dufferin, considered that it would be more useful for it to take on a political character as a sounding board for Indian opinion. Moreover, Congress tended to avoid discussions of women's status, which might antagonise some influential members. Only with Gandhi's rise to power did social issues become an integral part of its platform.

In the villages where 90 per cent of the population lived, these developments, except for the abolition of *suttee*, had hardly affected the lives of women. In the Non-Cooperation Movement which Gandhi launched in 1920 women at all levels were to be involved for the first time in ways which were to have a profound effect on their social position. Gandhi spoke of the movement for independence when he addressed the Gujerati Political Conference in 1917 in lyrical terms:

> The splendour of the spring is reflected in every tree, the whole earth is filled with the freshness of youth. Similarly when swaraj is upon us a stranger suddenly in our midst will observe the freshness of youth in every walk of life and find servants of the people engaged each according to his abilities in all manner of public activities. If progress has not been what it might have been, one reason is that we have kept our women away from these activities of ours and have thus become victims of a kind of paralysis. The nation walks with one leg only.[5]

In 1920 he launched his Non-Cooperation Movement, which called for withdrawal from government service and from government schools and colleges, refusal to pay taxes, and the boycott of foreign goods. Its symbol was the spinning wheel, with which its supporters were expected to make the cloth for their own clothes (*khaddar*). He turned to the women to help to establish *Ramrajya* – the Kingdom of God – in which women could live in safety and the starving millions see the end of hunger; in which there would be harmony between religious communities, untouchability would disappear, the cow would be protected and Indian culture returned to its proper place. The patience of women, he said, held the key to *swaraj*. He appealed to them to come forward and picket liquor shops and those which sold foreign cloth. He challenged them to love Hindus and Muslims equally and to treat untouchables as brothers and sisters. He urged them to spin cotton every day and to dress in homespun cloth. He asked them to give their jewellery to be sold for the cause and to pile up their costly saris and set fire to them like plagueinfected garments.[6] He called on them to participate in not only the political but the social movement. They must go out in the villages to revive spinning and preach fraternisation between Hindus and Muslims, between high castes and low castes. They must promote hygiene and literacy. Only women, organised by themselves, could enter the homes of women and change their attitudes; through such work great service would be rendered to Home Rule without the phrase being so much as mentioned.

His appeal was brilliantly successful. Women had been given tasks which at this stage were not likely to be dangerous or appear immoral and some of which could be carried out at home. Their husbands and fathers were reassured by the non-violence of their activities. The women themselves were inspired by the call for the exercise of the traditional Hindu feminine qualities of courage, patience and endurance. For most of them their jewellery was the only possession of which they had control so that they could feel a personal commitment in its sacrifice for the cause. Even those who stayed at home became supporters by spinning, at which they were much more proficient than men. To wear *khaddar* came to mean many things – opposition to foreign rule, and identification with the poor and oppressed, as well as the assertion of self-reliance and of freedom.[7]

In the Civil Disobedience Movement of 1930 they obtained a more active role. Gandhi had not wished them to join in his march to the sea to break the salt laws. He reflected (erroneously as it turned out) that Englishmen would not touch women, just as Hindus would not kill cows; it did not seem fair therefore for women to shelter behind men. He could exclude them from the march but he could not keep them from the sea. They came in their thousands in the symbolic breaking of the law by which the manufacture of salt was a government monopoly and its sale was taxed. Here was an issue which affected every kitchen, except, ironically, that of Gandhi himself who had renounced the use of salt as part of his strict dietary regime. In an atmosphere like a marriage festival the women filled their pots with sea water, returned to their homes to make salt and then sold it in the streets for the cause. Before Gandhi was arrested he accepted the women's insistence on an active role and nominated Mrs Sarojini Naidu to lead a raid on a government salt works. The women were beaten by the police, and press photographs of their brutal treatment obtained wide publicity for the movement abroad.[8]

The Congress leaders in the movement of 1930 were rapidly arrested, 'and then', wrote Jawaharlal Nehru, who was one of them,

> a remarkable thing happened. Our women came to the front and took charge of the struggle.... There was an avalanche of them who took not only the British Government but their own menfolk by surprise. Here were these women, women of the upper or middle classes, leading sheltered lives in their homes – peasant women, working class women, rich women – pouring out in their tens of thousands in defiance of government order and police *lathis* (batons). It was not only that display of courage and daring but what was even more surprising was the organisational power they showed.[9]

In the salt movement alone 17,000 women were arrested. In response, Congress at its Karachi session in 1931 thanked the women for their part in the civil disobedience campaign and pledged itself at Independence to sexual as well as caste and religious equality before the law, to the abolition of discrimination in public offices, and to universal suffrage.

Part of Gandhi's appeal to uneducated women lay in a homely style of speech and writing not unlike that of the parables of the Christian New Testament which he so much admired. He told them, when he asked them to give up their jewellery to his campaign, that ear and nose rings were instruments of slavery by which they could be led by men. When they complained that homespun cloth was coarse and unattractive he replied that he had never known a woman to throw away her baby because it was said to be ugly. Discussing the roles of men and women in the independence movement he would remind them how a cart would cease to run properly unless both wheels were in order. He explained that fear is a sign of lack of faith. They were, he told them, by nature better as Satyagrahis than men. They must put their trust in God to protect them. For a river, he pointed out, is ever ready to give water to all: but if one does not approach it with a pot in which to fetch water, or avoids it, thinking it poisonous, how can that be the fault of the river? In their picketing he urged them to go among drunken men as fearlessly as he had seen Salvation Army girls go into the dens of thieves and gamblers and drunkards in England. He advised them to cut off their hair so that they could not be dragged by it. His message was that women were not weak, and he held up as models the legendary Hindu heroines such as Sita and Draupati, as well as Joan of Arc, the supreme example of chastity, who inspired the French to drive the English out of their country.[10]

For many women part of the attraction of the movement was that it enabled them to escape from the boredom of domestic life. Some of them looked back to participation as the happiest time of their life. Mrs V. L. Pandit in her memoirs recalled how Gandhi's magic had caused women to leave the shelter of their homes and, by taking part in the Non-Cooperation Movement, achieve an equality denied by custom. Yet she added,

> I have never quite forgiven myself for that first jail sentence which broke up my home when my children most needed its security and comfort. To stay at home and look after them would have been dull. Whatever the reason, I am now sure that I acted selfishly thinking in vague terms of personal political achievement rather than the satisfaction I could have gained through domestic duties honestly performed.[11]

For considerable periods Gandhi withdrew from politics and gave most of his attention to his constructive programme. In this, education had an important part. Whilst the way to freedom did not lie through education, he taught, education must follow freedom. Without it mankind remained animals. It was not justifiable, however, to withhold rights meanwhile on grounds of illiteracy. His views on the education of girls evolved as they changed regarding the employment of women. He started with the firm Victorian idea that the woman's place is the home. 'A state of affairs in which women have to work as telegraph clerks, typists or compositors can I think be no good; such a people must be bankrupt and living on their capital,' he told a meeting of mill hands in 1920. 'Women should not work in factories. They have plenty of work at home. They should attend to the bringing up of children. They may give peace to the husband when he returns home tired, minister to him, soothe him if he is angry and do any other work they can staying at home.'[12]

By 1931 he admitted, however, in an interview with a British newspaper, 'I want to see opening of all offices, professions and employment to women. Otherwise there can be no equality. But I most sincerely hope that woman will retain and exercise her prerogative as queen of the household. I cannot imagine a happy home in which the wife is a typist scarcely ever in it.'[13] By 1940 he conceded, 'All will work according to capacity for an adequate return for their labours. Women in the new order will be part time workers, their primary function being to look after their homes.'[14]

He urged Congressmen to educate their wives, daughters and even mothers in their homes but taught that whilst men were responsible for the defects in women, in the end the latter's regeneration could only come through self-help. In 1924 his preoccupation with *khaddar* led him to say that it was sufficient education for girls to learn how to spin. A few years later however, he indicated that women should learn Hindi, if it was not their mother tongue, enough Sanskrit to understand the Bhagavadgita, elementary arithmetic and composition, music and child care, all of course in addition to spinning and weaving – a substantial curriculum. As to how far girls should have the same education as boys and be taught together with them his views also varied. In general he favoured co-education at the primary level and a separate education afterwards in which that of girls should place an emphasis on domestic arts.[15] Some of his early pronouncements on co-education however were conservative, even puritanical. At the secondary level, he said in 1920, boys and girls should sit in separate rows, in order that there should be no personal contact. They should not be allowed to correspond or joke with each other. Later, as they grew up, he said that it was sinful and reprehensible for a man to seek a woman's company for its own sake and desirable that a man and woman should avoid being alone together. This stern position, however, was relaxed in his ashram, as it came to be understood that its members regarded each other as brothers and sisters.[16]

Gandhi had called women out of their homes to participate in and even lead the Non-Cooperation Movement; it could hardly be expected that they would all return afterwards to be only housewives. It would have been inconsistent to compare Indian women unfavourably with the British in their failure to come forward as nurses and social workers and still to urge them all to stay in their homes. Out of the movement therefore developed the concept of the career of the unmarried woman social worker, working mainly in the villages and just as much honoured as the wife and mother. When Congress came into power women became legislators and even ministers. The suffrage indeed was never a problem. As Gandhi told a meeting of British women in London in

1931: 'You had to suffer for suffrage. In India women got it for asking.'[17] Indian women, he said, had set a fine example by not claiming privileges. Congress under the leadership of Gandhi and Nehru was so committed to the political equality of women that this never needed to become a campaigning issue.

Women, Gandhi said, were in the position of a slave who did not know how he ever could be free, and who, when freedom came, for the moment felt helpless. It was up to Congressmen to realise their full status and enable them to play their parts as equals of men.[18] The equality of women with men, he maintained, would only be achieved when the birth of a girl was celebrated with as much joy as that of a boy. Such a change in social attitudes faced formidable obstacles in the Hindu scriptures. Pandita Ramabai, in the first book by an Indian woman to be published in the English language (*The Hindu High Caste Woman*, 1887), had quoted indignantly the laws of Manu, which laid down that a wife must worship her husband as a god even if he was seeking his pleasure elsewhere or was devoid of good qualities. In her revolt against this she had moved from Hinduism to Christianity. Gandhi's response to the problem was to insist that all the laws in the Hindu scriptures which were demeaning to women were later insertions which should be disregarded.

Assertion of the principle of the equality of women often took Gandhi into discussion of their place in the family. Purdah he described as a relatively recent institution, introduced into India by foreign invaders; it created moral weakness in women. He deplored the ostracism of widows, but, whilst urging that child widows should be allowed to marry, greatly admired the devout and retired life chosen by mature widows. Though he himself had been married at 13, he regarded child marriages as disgusting and in his own case he always felt guilty that he had not given sufficient attention to the education of his wife. On property he taught that a daughter's share should be the same as that of a son and that husband and wife have equal rights in what each other earns. He considered it better for a girl to remain unmarried rather than to be humiliated by having to marry a man who demanded a dowry. He criticised the lavish expenditure felt to be necessary on weddings, which often left crippling, lifelong debts. Simple 'Gandhi weddings' thus came to be practised instead. He showed little interest in proposing legislation to bring about equality between the sexes but much in arousing the change of heart which must precede it. He told Congressmen that the freedom movement should begin in their homes by the removal of all restricting customs and conventions. If they would not allow their wives and daughters to go out and canvass he said, 'Send the names of such antediluvian fossils to me for publication in *Harijan*.'[19] Whilst he did not as a rule encourage the wife to follow a vocation independent from her husband, on the other hand he said that every woman should take up some form of work. As he came to recognise the expanding role of the women workers he asked, 'If a woman is not allowed freedom from household chores, or if she wastes her time in decking herself up and gossiping, what service can she do for her country? In my opinion the slavery of the kitchen is a remnant of barbarism.'[20]

No one, it has been suggested, least of all Congress, knew what Gandhi would say on a particular question of social reform.[21] His position on questions of sexual conduct could appear bizarre and even cranky. Throughout his life he was implacably opposed to artificial methods of birth control, which he called as good as death, putting a premium on vice and resulting in imbecility and nervous prostration. Between him and Margaret Sanger, the President of the American Birth Control League, who came to see him in 1935, or Lord Dawson of Penn, whose eloquent address in favour of family planning was sent to him, there was no agreement whatever on fundamentals. Mrs Sanger regarded sexual expression as a physical need for women as well as men. Gandhi saw it to be a very occasional necessity only justifiable in order to propagate the race. When Mrs Sanger asked if sexual union therefore might thus only take place three or four times in a woman's lifetime, he replied that people should be taught that it was immoral to have more than three or four children and that after that they should sleep separately; if social reformers could not impress this idea there should be a law on the matter. He described Lord Dawson as biased in indicating that sexual repression could be harmful, because doctors only dealt with unhealthy rather than normal people. When Mrs Sanger persisted, he did concede however that in some circumstances the safe period technique might be permissible.[22]

Although Gandhi had once maintained that seeking women's company for its own sake was sinful and reprehensible he came to appreciate that association between the sexes in public work was inevitable and he enjoyed the company of female helpers such as the privileged jester Sarojini Naidu. In middle age he had a close, though of course unphysical, relationship with a married woman, Sariandevi Chaudran, who he asked to be his 'spiritual wife', somewhat to her embarrassment. A number of women have left accounts of Gandhi's personal impact on them. When Sarojini Naidu first met him she was already a leading advocate of women's suffrage who had studied at Cambridge and was a celebrated poet. Climbing up the many stairs of the unfashionable lodging house where he was staying whilst on a visit to London in 1914, she found him seated on his black prison blanket, eating a messy meal of squashed potatoes and olive oil; around him were battered tins of groundnuts, tasteless biscuits and dried plantain flour. 'Come in and share my meal,' he invited her. 'What an abominable mess is this!' she replied.[23] She became his devoted disciple, taking a leading part in his campaigns and loyally going to jail. She teased Gandhi, calling him 'Mickey Mouse' and telling him that he did not know what it cost to keep him in poverty. When expected to spin, she said that her thumb ached. Swadeshi for her meant not so much spinning as the revival of each Indian art and craft, a renaissance in literature and a new vision of architecture. Fighting for women's freedom and self-respect she found fully consistent with being feminine and she continued to wear gorgeous silk saris instead of homespun ones because they cheered people up. When Gandhi was assassinated she said to her fellow disciples, 'What is all this snivelling about? Would you rather he died of decrepit old age or indigestion? This was the only death great enough for him.'[24]

Very different was Gandhi's relationship with the British admiral's daughter Madeleine Slade, who became Mira Behn when at the age of 32 she joined his ashram. She immediately insisted, almost against Gandhi's advice, in cutting off her hair and adopting Indian dress. From early morning to last thing at night she lived for the moments when she could set eyes on him. As she sought to crush her natural independence of nature and put herself under his will he protested, 'You must not cling to me. You must retain your individuality.' He became fidgety and it seemed best for her to work away from him. 'You have left your home, your people and all that people prize the most,' he wrote, 'not to serve me personally but the cause I stand for. All the time you were squandering your love on me personally I felt guilty of misappropriation and I exploded on the slightest pretext.' There were times when she was able to be close to him. It was useful for him to have her at his side when he was in London for the Round Table talks. But latterly she ran her own village uplift schemes. After his death she returned to live in Vienna where she devoted herself to the study of Beethoven and disliked any discussion of that half of her life spent as his disciple. It is a somewhat sad story.[25]

Gandhi's wife Kasturba left no memoir, being almost illiterate, but in his own autobiography he vividly describes her indignation when in the South African ashram she was required to do the sweeper's job of emptying chamber pots. She was initially unhappy at having to live with Harijans and share meals with all castes. Gandhi says that she agreed willingly to cease having sexual relations. 'Kasturba', he wrote, 'herself does not perhaps know whether she has any ideals independently of me. It is likely that many of my doings have not her approval even today. We never discuss them. I see no good in discussing them. For she was educated neither by her parents nor by me when I ought to have done it.' Yet he added that when they disagreed on practical matters, 'my wife with her matchless power of endurance is always the victor'. It has been said that Kasturba could make him feel absurd just by speaking one devastating home truth, for which she had a genius.[26]

How far Gandhi can be regarded as sympathetic to feminism has often been disputed since his death. Women ought to learn to live together, he said in 1929, to work together, to tolerate one another's temperamental differences, to think independently and to put these thoughts into action with courage and determination; however, the women of India had not yet developed a viewpoint that enabled them to look beyond their families.[27] He appealed to students to emulate the teaching of Browning, Whittier and, surprisingly, Milton, and treat women as queens of their hearts and homes and not slaves: but on the other hand he regarded the western, purely formal, yielding of first place to women as highly artificial and even hypocritical.[28] Although Gandhi honoured and drew inspiration from other religions, notably at his daily prayer meetings, he was always a devout Hindu. When he addressed mass meetings he would often hold up the example of the goddess Sita for emulation, enduring grim hardships in following her husband and resisting seduction by a demon, and yet having an independent spirit. She was pressed into serving the *Swadeshi* campaign as an example, in having clothed herself in the barks of trees. Spinning he constantly urged on women as a religious duty.

Despite the validity of some criticisms by feminists after his death, Gandhi's achievements for women were of great importance. Unlike liberal predecessors he did not regard women as objects of reform but as active agents of social change. He saw their situation as inextricably linked to wider issues. Liberation of women, liberation of India, removal of untouchability, amelioration of the economic condition of the masses resolve themselves into penetration of villages and reformation of village life, he would say, and then in the process of reconstruction women would free themselves from bondage.[29]

He created a tradition by which hardly anyone later would be able to stand up and explicitly oppose women's rights or deny them participation in politics. Gandhi's teaching on women's rights might however have been much less effective if they had not been fully shared by Nehru from the point of view of a socialist. After independence India's legislators found little difficulty in granting women political equality or even in admitting them to the elite Indian Administrative Service, the successor to the Indian Civil Service, the 'steel frame', as Lloyd George called it, which had held the Indian Empire together. Equal social status encountered much more resistance and it was only the great prestige of Nehru as Prime Minister which finally enabled legislation to be passed which gave women rights to succession to property and in marriage and divorce.

Towards the end of his life Gandhi said, 'I have radical views about the emancipation of women from their fetters which they mistake for adornment. I hope one day to place some of my conclusions before the public when my researches are completed.' He was murdered before he had the opportunity to do so.[30] It would be imprudent to try to set down general conclusions about his views on women – or indeed on other questions – because his experiments with truth never ceased. He could be very dogmatic on some points, as Mrs Sanger experienced; but on others he would say, 'Never take anything as Gospel truth even from a Mahatma,' or, 'Correct me if I am wrong,' when even the youngest of us questioned him. How indeed this elderly man, burdened with his country's appalling problems in the run up to Independence, found time and energy to enter into the personal concerns of everyone who consulted him seemed almost miraculous. Sometimes as he cross-questioned us in return there was a suspicion that he was teasing the self-confidence of youth.

Whilst Gandhi was a brilliantly imaginative politician, politics were only an occasional preoccupation of a moral teacher whose parables were sometimes as much intended to make his listeners think out their position for themselves as to be taken literally. Though he was constantly concerned to improve the position of women, he also sought to stimulate them to work out for themselves what their role should be in the new India. The substantial volume of writing by Indian women which has continued since his death on the issues which he raised, though sometimes highly critical, particularly of his neglect of the economic position of women, is perhaps in this broader sense an indication of his success.

Part Two
The Assault on the Citadel

6 Education

How can you give a woman self respect?

F. D. Maurice, Inaugural Lecture at
Queen's College, London (1848)

The men who began to work for improved secondary education and access to higher education for girls and women in the mid-nineteenth century did so for several motives. Perhaps the most widespread, now that success in the professions was no longer barred by class but was being opened to competition, was that educated mothers could give their sons a grounding and a start in life. Then there were men who reflected that in the long run educated women are likely to be less boring marriage partners than the ignorant. Educated daughters too might be able to earn a little money and relieve the expenses of large middle-class Victorian families, though there were few openings for their employment except as governesses and companions, and in writing.

Such were the practical motives. But there were also men who took up the cause of female education because they sympathised with the despair of intelligent women whose abilities found no outlet – the despair so vividly illustrated in the early letters of Florence Nightingale in the long years before she was allowed to pursue her vocation. A few, such as J. S. Mill and, earlier, Sydney Smith, deplored the waste of half the nation's potential talent. The latter in an article in the *Edinburgh Review* in 1810 had protested against the general belief that the education of women and girls would lead to neglect of domestic duties. 'Can anything be more absurd', he asked, 'than to suppose that a mother's solicitude depends upon her ignorance of Latin or Greek or that she would desert an infant for a quadratic equation?' Of those selected for consideration in this chapter F. D. Maurice was an idealistic Church of England clergyman bearing the imprint of a Unitarian background. George Grote and Henry Morley saw the development of women's education as part of London University's general liberalising mission. Thomas Holloway was a childless millionaire who happened to make this the object of his philosophy. Henry Sidgwick wanted women to be admitted to Cambridge University partly because this would help in modernisation of its curriculum. There was no disagreement among them that the establishment of a sound foundation of secondary education for girls was the first necessary step. On the nature of the next stage, higher education, there were uncertainties and differing views – should men and women study together or in separate institutions? Should the curriculum for women be the same as that for men or was something different required? How important was it for them to obtain degrees rather than accept certificates of their studies? And if they were granted degrees how necessary was it to insist that they have a share in the governance of universities and thus antagonise conservatives whose cooperation in teaching them was essential? Cardinal Newman maintained in his influential book *The Idea of a University* (1852) that young men gained more from open-hearted, free-ranging discussions between themselves than they did from instruction by their tutors. But such bold exchanges between young men and women were generally precluded by Victorian taboos and conventions of chaperonage. Indeed both men and women students felt inhibited from initiating them. At the Homerton Congregationalist Training College, for example, which consisted of separate male and female institutions, the Principal's attempt to organise joint literary and social gatherings failed because the women said they could not speak freely before men. Correspondingly, a scientist as sympathetic as Huxley felt that there could not be frank discussion at meetings of learned societies at which women were present. It could be argued that, at least in what might be hoped to be a transitional stage before Victorian prudery was overcome, women could educate each other along Newman's lines in forthright debates within their own colleges and societies better than in bland discussions when the sexes met together. Yet on the other hand by doing so they would miss the stimulus of the most brilliant male contemporary minds. The men who took the lead in the movement for women's education were constantly preoccupied with these issues.[1]

A PROPHET – F. D. MAURICE

When Queen's College, London, celebrated its centenary in 1948 Sir Frederick Maurice, the grandson of its founder, wrote proudly,

> 1848 is known to historians as the year of revolutions, but historians have hardly noticed the fact that in that year there began in England the greatest revolution of them all. It was a revolution which has profoundly affected the national and social life not only of our country but of the world at large; for in that year was opened the first college for the higher education of women.[2]

The immediate inspiration for the involvement of F. D. Maurice and his friends in the cause of education of women and girls came from their concern for the situation of governesses, whose overcrowded profession was the only one open to any extent to middle-class women. Maurice was drawn by his sister, who was a teacher, into supporting the Governesses Benevolent Institute. At its inauguration Charles Dickens spoke passionately of governesses who were paid worse than the cook and 'compared but shabbily with the remuneration of the ladies' maid'. It was time, he urged, to blot out a national reproach and for the educational profession to be placed on the honourable footing which in any civilised and Christian land it ought to hold.[3] Maurice and his associates, mostly clergymen and university teachers, started by setting up an examination to test and provide a certificate for governesses which would raise their status. The results however showed such a degree of ignorance that they felt bound to go further and establish a college to educate them before they were examined.

Frederick Dennison Maurice (1805–72) was one of the most interesting of nineteenth-century theologians. His prophetic insistence on the social gospel of the Church at a time when the principles of laissez faire were dominant was unusual in its time: his influence indeed was perhaps less important directly than afterwards through such disciples as Bishop Westcott, Stewart Headlam and Octavia Hill. His thinking was influenced by his upbringing, as son of a Unitarian minister. He had eight sisters, 'none of them', as one of his biographers dryly comments, 'of a retiring disposition'.[4] Maurice described his mother as having a far clearer intellect and much more lively imagination than his father. The women of the family had a keen interest in religious dogma which they would communicate in writing, strongly attacking each other's beliefs. Mrs Maurice and several of the sisters became Calvinists and ceased to attend the church of their husband and father, much to his grief. Later several of the sisters joined the Church of England and one became a Baptist.

Growing up in this atmosphere, Maurice developed two principles which were to be of great importance throughout his life, firstly a belief in the intellectual capacity of women, and secondly a desire for a spirit of unity in which members of different religious denominations and groupings would try to appreciate the best in the beliefs of others instead of attacking them. As an undergraduate at Trinity College, Cambridge, he became a leading member of the Apostles Society, among whose principal interests was the education of women. One of his fellow members, Alfred Tennyson, as has been seen, was later to establish his reputation with his long descriptive poem *The Princess* about an educational academy for women. Another Apostle, Monckton Milnes, almost persuaded Florence Nightingale to marry him. Maurice, in the *Metropolitan Quarterly Magazine*, which these students launched, strongly attacked the current system of young ladies' education. Instead of having their minds broadened, he said, they were taught by means of little catechisms of facts, dates, the nomenclature of chemistry and astronomy and so on, with no consideration for the causes of historical change or the knowledge of scientific principles. 'The imagination', he asserted, 'is a terrible object of the dread, the hatred and hostility of the mistresses of establishments and the governesses of young ladies.'[5]

In his twenties Maurice was ordained in the Church of England and at the age of 35 he was appointed Professor of English Literature and Modern History at King's College, London, which had recently been established by Anglicans as a response to the 'godless' University College, London, many of whose founding fathers were Unitarians. At the same time he was Chaplain of Lincoln's Inn where his sermons attracted people of all denominations.

It was now that Charles Kingsley drew him into the Christian Socialist movement and that Maurice's great desire became 'to Christianise socialism'. 'Competition', he declared, 'is put forward as the law of the universe. The time is come for us to declare it is a lie by word and deed.'[6] He and his friends enthusiastically, but not very successfully, organised cooperative societies for working men and women as a protest against the assumption that selfishness was the basis of society.

It was in 1847 that Maurice persuaded a number of his fellow professors at King's College to form a committee to examine and teach governesses. Queen Victoria was persuaded to grant a royal charter; thus Queen's College was established, which very soon became open not only to governesses but to other women and girls. Except for the language teachers all the professors were Anglican clergymen, mostly from King's College and working on a part-time and voluntary basis. One of the most influential was Charles Kingsley.

Maurice was Principal from 1848 to 1853. He used his inaugural address to express ideas on the education of girls which were an expansion of those which he had asserted as an undergraduate in Cambridge. After describing the wretched state of governesses he raised the question, 'How can you give a woman self respect?' Praising the deep wisdom of Tennyson's *Princess*, he insisted that the new institution was to be a college, not a school: for whilst teachers in a school may aim merely to impart knowledge, those in a college must lead pupils to a study of principles, by which he meant 'something as far as possible from introducing them to an encyclopaedia of knowledge' or to teaching them 'accomplishments'. In our eagerness to 'finish' girls, he observed, we are not equally solicitous about beginning. 'To know a single fact is a blessing unspeakable; to know *about* a thousand rather a perplexity and torment.' He applied these principles through the subjects to be studied. Thus music, he said, awakened a sense of order and harmony in the heart of things which outwardly were most turbulent and confused. Drawing encouraged a habit of observation of form, a power of looking below the surface of things for the meaning which they express. In mathematics we can come to regard numbers with the kind of wonder with which a child regards them, to feel we are looking into the very laws of the universe. When grammar is regarded as a meditation upon roots and growth of words, children's faces become brighter and fuller of meaning as they engage in it. English literature must not be taught by books about books: Shakespeare and Milton are the best critics of themselves. History and geography should be studied not because they have a meaning to draw out of them but because we must put a meaning into them. The approach to theology would be historical. Above all, women must not study in order to be admired.[7] It was a noble vision.

The association with it of Maurice and Kingsley led to attacks on Queen's College for teaching 'a sort of modified pantheism and latitudinarianism'. The list of lecturers was so impressive however that such criticisms carried little weight. They included J. R. Green, the most popular historian of his time, George Richmond, the most fashionable portrait painter, Dean Stanley and Stopford Brooke. The College educated many of the pioneers in the movements for equality of women, including the great headmistresses, Miss Buss and Miss Beale as well as Sophia Jex-Blake and Louisa Twining. Yet it never became a university college as it could have done after women were admitted to full membership of London University. Most of the parents of these upper-class London girls saw a broad secondary education for their daughters as an asset in the marriage market, on which they were usually launched at the age of 18. If their education, however, continued to 21 or 22 it was feared that they might have lost their bloom and their innocent acceptance of male intellectual superiority. For the minority who wanted higher education University College, London, with its far larger staff and facilities, was more attractive, after it opened its doors to women in 1878, than Queen's could be. For those who wished Queen's to become a university college there was an inherent weakness in the founders' decision to admit girls from the age of 12. Despite Maurice's distinction between a college and a school in his inaugural lecture, Queen's settled down to become an excellent small independent single-sex school, still, as in Maurice's time, setting out to discover and develop the individual gifts and aptitudes of each person, still remembering Maurice's assertion that competition was not the law of the universe.[8]

In 1866 Maurice was elected Professor of Moral Theology at Cambridge. Here he remained for the rest of his life and his lectures were attended by as many women from outside the University as by men. From his early experience of the talents shown by Unitarian women and through teaching Greek to his sisters he had become convinced of women's intellectual capacity. He had cast the net of Queen's College beyond potential governesses because he saw 'every lady as a teacher of some persons or another – children, sisters or the poor'.[9] He supported Mill's proposal for votes for women as a positive strengthening of the moral life of Britain, considering that in any sphere in which women feel their responsibility they are as a rule more conscientious than men. He championed the Married Women's Property Act because, he said, the ethos of marriage is trust.[10] He regarded history as a process of progressive divine revelation, of which the emancipation of women was a part. As he lay dying his family made out his last words to be 'the Communion is for all nations and peoples' and 'something too, we understood, about it being *women's* work to teach men its meaning'.[11]

GEORGE GROTE AND HENRY MORLEY AT LONDON UNIVERSITY

It was London University which in 1878 became the first in Britain to admit women to its degrees. That it did so was a natural development, for it had been founded in 1835 mainly by Radicals and Nonconformists determined to open its doors more widely than those of Oxford and Cambridge. George Grote (1794–1871), who first raised the issue in the Senate, was the University's Vice Chancellor and later President of one of its most important constituent bodies, the University College of London (UCL). A friend of Bentham and James Mill, he had been a reforming MP until he decided to devote himself to writing his massive *History of Greece*. He was strengthened in his feminism by a formidable wife whose biographer says that 'he formed her mind with a solid basis; she fashioned, moulded, framed and glazed him'.[12]

In 1862 the University received an application from Elizabeth Garrett's father for her to be examined in medicine. Legal advice was

1. George Meredith, portrait by G. F. Watts, 1893, National Portrait Gallery. His novels sympathised with the predicament of women struggling against convention.

2. George Bernard Shaw, water-colour by Bernard Partridge, National Portrait Gallery. His plays urging women to be unwomanly had a wide influence because they were so entertaining.

3. John Stuart Mill by 'Spy', *Vanity Fair*, 5 April 1873, 'A Feminine Philosopher'. He was the first MP to propose women's suffrage and his book *The Subjection of Women* became the bible of the feminist movement.

4. Jacob Bright by 'Spy', *Vanity Fair*, 5 May 1877, 'The Apostle to the Woman'. He took over the leadership in Parliament of the women's cause after Mill lost his seat.

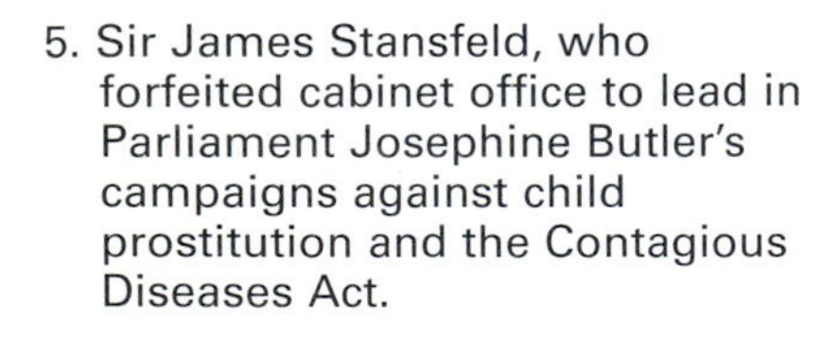

5. Sir James Stansfeld, who forfeited cabinet office to lead in Parliament Josephine Butler's campaigns against child prostitution and the Contagious Diseases Act.

6. Lord Dawson of Penn, who challenged the opposition of the Church and the medical profession to contraception.

7. F. D. Maurice, chalk drawing by Samuel Laurence (*c.*1846), National Portrait Gallery. Founder of Queen's College, London.

8. Thomas and Jane Holloway, statue at Royal Holloway College (Royal Holloway archives PH118). Inspired by his wife, the millionaire Holloway built a magnificent women's college modelled on the châteaux of the Loire.

9. General William Booth, cartoon in *St Stephen's Review*, 20 February 1892. He caused a sensation by giving women command over men in the Salvation Army.

10. Bishop R. O. Hall of Hong Kong, who controversially ordained the first woman priest in the Anglican Church in 1944.

11. Sir Henry Acland, who was instrumental in causing the medical profession and Queen Victoria to cease to oppose the training of women as doctors.

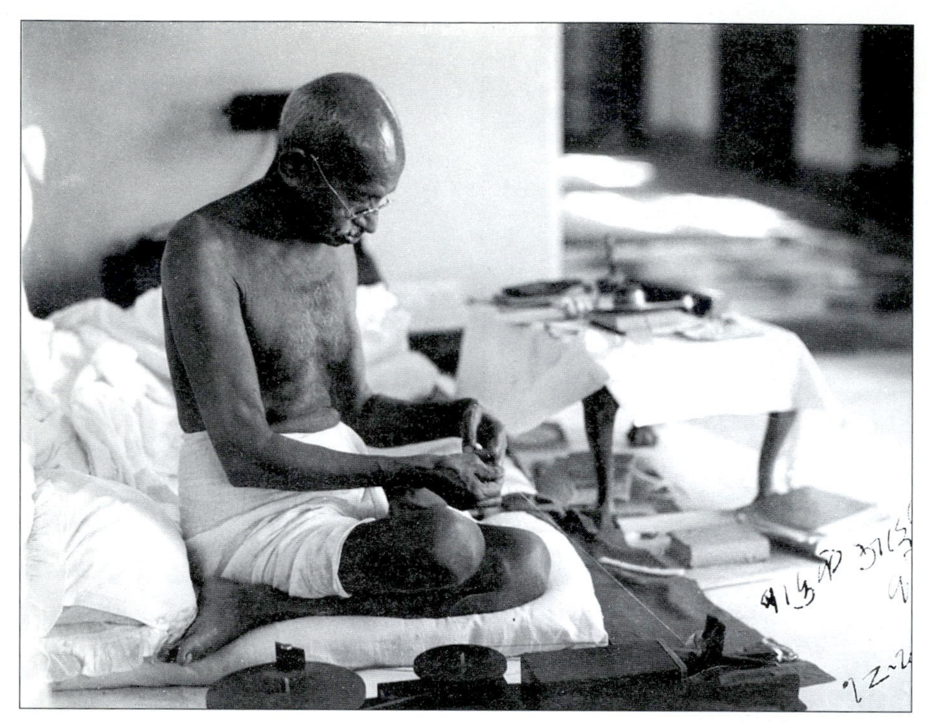

12. Mahatma Gandhi. He brought Indian women into public life through the freedom movement (author's own photograph).

13. Benjamin Disraeli by J. E. Millais, 1881, National Portrait Gallery. As leader of the Conservative Party he told Parliament that there was no reason why women should not have the vote.

14. James Keir Hardie, *Vanity Fair*, 8 February 1906. He threatened to resign the leadership of the Parliamentary Labour Party unless it gave priority to women's suffrage.

15. David Lloyd George by Sir William Orpen, 1927, National Portrait Gallery. He was Liberal Prime Minister of the coalition government which granted women the vote in 1918.

16. Henry, Lord Brougham by James Lonsdale, 1821, National Portrait Gallery. He was influential in enabling married women to own property and obtain divorce.

17. 'The Judgement of Parisette'. Cartoon in *Punch*, 27 December 1911. It depicts Richard Haldane, Sir Edward Grey and Lloyd George competing for the championship of the women's cause.

received that this was not permissible under the University's charter. Grote moved in the Senate that it should seek to modify the charter, whose original object had been advancement of education among all classes and denominations without any distinction whatever. No one in approving it, he argued, could have imagined that it would exclude half Her Majesty's subjects and all of her own sex. Much teaching was now being done by women, and not only they but the public would benefit if their qualifications could be determined by examination. There was also a larger proportion of women than ever before who were dissatisfied with a life of dependence without work, an essential preliminary to which was the cultivation of habits of application. Women's exclusion from the University was unfair and objectionable.[13] Some years before J. S. Mill's *The Subjection of Women* was published this was a startling speech but it almost carried the day. The Senate's vote was a tie, broken by the casting vote of the Chancellor, Lord Granville, in favour of the status quo. Yet though the Citadel had survived, its foundations were shaken. Three times the University Convocation passed resolutions in favour of permitting women to take degrees. Eventually the Senate sought and obtained a supplementary charter in 1877 which permitted them to do so.

The University, however, was only an examining body whose component colleges were responsible for teaching. In 1869 UCL obtained a revised constitution which allowed women to be admitted. The moving spirit was Henry Morley (1822–94), whose feminism, like that of George Meredith, seems to have been inspired when he studied as a boy with the Moravian Brothers in Germany. He became in turn a surgeon, school proprietor and lecturer in English at King's College, London. Debarred from obtaining a chair there because he was a Unitarian, he became Professor of English Literature and Languages at UCL. He was the most prolific populariser of English literature of his time.

Morley felt strongly that 'it was the business of UCL to be boldly first in recognising fully any new and real want of the time'. One such want, he believed, was for men and women to be able to study together. Unlike Grote he avoided raising a question of principle. He started by causing a Ladies' Education Association to be formed with a committee of ten men and three ladies in order, he said, to gain the confidence of prospective students that through it 'feminine requirements would be properly considered'. In 1869, the year in which the Association was established, *The Times* declared in a leading article 'we doubt whether a system under which young women should receive exactly the same sort of education as that now given to young men, the two sexes mixing in the same classes, pursuing the same studies and contending for the same degrees, will ever prove desirable in this country'. 'Despite this warning,' Morley later recollected, 'we pushed boldly on, encouraged by the continuing advance of liberal opinion in the matter of education of women.' Pleading the Association's poverty, he obtained the use of the College's rooms and equipment for women's classes, each commencing half an hour after those of the men started and with entrance to the building by a separate door so that there was no contact between the sexes. He persuaded College professors to give lectures to the women in physics, chemistry and philosophy whilst he undertook those in English Literature. In 1869 the College obtained a modification to its charter to enable it to teach women in all subjects except medicine, though they could not obtain degrees until the University itself altered its charter. The next step was to establish a precedent for holding mixed classes on the practical grounds that Professor S. E. Cairnes, who taught Political Economy, was an invalid who could not be expected to teach men and women at different times in separate rooms. Gradually mixed classes in other subjects were quietly arranged. Only E. S. Beesly, Professor of History and one of Britain's leading Positivists, protested against the mixed classes and later against the appointment of women teachers. Morley's final contribution was to set up and preside over a hall of residence for the women at UCL. By 1878 when women were admitted to degrees, 320 were already studying at the College. The Ladies' Association could now be disbanded. By the late nineteenth century a third of UCL's students were women, a far greater number than at Oxford and Cambridge together. There were many more in the separate women's colleges now established within London University, and all the new civic universities which were founded in the late nineteenth century admitted women to full membership. *The Times* on this as on many other aspects of women's emancipation had proved a poor prophet.[14]

A PHILANTHROPIST – THOMAS HOLLOWAY

The Royal Holloway College for Women, which opened in 1886, was planned and financed by a millionaire who had made his fortune in selling patent medicines: in his initiative he was emulating and competing with the example of Matthew Vassar, the founder of Vassar College in Poughkeepsie, New York. 'My ambition', said Holloway, 'is to leave the

College so complete that its equal cannot be found in Europe or America.... I hope I shall be able to beat Vassar into fits. It shan't be my fault if we don't.'[15]

Thomas Holloway (1800–83) was the son of a retired sea captain in Cornwall who ran a bakery and an inn. In his teens he had been apprenticed to a druggist and this experience was to enable him to make his fortune. From 1839 he manufactured pills and ointment and spent huge sums on advertising them as a cure for every kind of ailment. He once went bankrupt but recovered to become a millionaire. When he was in his sixties, having no children, he began to discuss with philanthropists such as Lord Shaftesbury and Gladstone how to dispose of his fortune. On their advice he founded an institution to provide treatment of mental diseases for those members of the middle classes who were unable to afford private treatment but were not poor enough to enter public asylums. It was much ahead of its time in care and treatment, and each inmate had their own room.

Prompted by his wife he then turned to the idea of founding a college – or university as he liked to call the project – for women. 'Most of us', he wrote to a friend, 'who do well are indebted in early youth to the teaching and training of mothers, and how much better it might be to the human family if mothers of the next generation should possess a high class education – and if any one took a degree would not boys and girls be formed of such a habit and say "Mother whatever you have done in the way of learning in the past, we will strive to emulate you".' The College, Holloway asserted, was not to be a training college for governesses but a ladies' university which would hold the same relation to higher education of women as Oxford and Cambridge did to men. The clerical influence, usually so prominent, was to be excluded; not only clergymen, but lawyers and doctors (few of whom perhaps were well disposed towards the patent medicine business) should be ineligible to serve as members of the governing body. Though he consulted closely with Emily Davies and Millicent Fawcett, women, too, were not to be governors. The curriculum was to include modern and classical languages, algebra, geometry, physics, drawing, and study of Mill's Logic, J. R. Green's English History and such other subjects as might be suggested by Professor Henry Fawcett and other competent advisers.[16]

'The founder believes', he said, 'that the education of women should not be exclusively regulated by the tradition and methods of former ages but that it should be formed on those studies and sciences which the experience of modern times has shown to be the most valuable and best adapted to meet the intellectual and social requirements of the students.'

The religious teaching was to be such as to impress most forcibly in the minds of the students their individual responsibility to God, whose hand in his life he had witnessed in all things.[17] The foundation deed expressed the hope that by Act of Parliament or Royal Charter the college would be enabled to grant degrees.

Whilst considering the curriculum and organisation of his college or university Holloway simultaneously worked enthusiastically on the architectural design. He selected his wife's brother-in-law, George Martin, as his successor to complete the work. With Martin and the chosen architect he visited and measured all the rooms in the principal châteaux of the Loire and selected that of Chambord as the model. He then made a similar tour of Cambridge to obtain inspiration from the layout of its colleges.

Holloway was not without vanity. Earlier in life he had signed his letters as 'Professor'; he designed for himself a coat of arms with a goat's head and *nil desperandum* under it; he gave instructions that there should be an imposing monument to his wife and himself in the front quadrangle. This small weakness was to have important consequences. When he died in 1883 the buildings were not completed. His brother-in-law took the name Martin-Holloway and vigorously continued the work, employing as sculptor of the founder's memorial Count Gleichen, the son of Queen Victoria's stepsister. Gleichen approached the Queen directly and persuaded her both to open the college in person and to allow it to take the title of 'Royal Holloway College'. Her courtiers were somewhat scandalised that she should have given her patronage to an educational institution in which the Church of England had no influence; the Queen's secretary told Martin-Holloway that before the college was opened suitable governors must be appointed.[18] Martin-Holloway was an ambitious snob. He was now easily persuaded to invite to become governors the Archbishop of Canterbury, as well as Randall Davidson, who was Dean of Windsor and a subsequent archbishop, together with Prince Christian of Schleswig Holstein and Sir Henry Thring, all of whom disapproved of the College's undenominational character.

The building which the Queen opened a year after Holloway's death was magnificent and bizarre. Years later John Betjeman was to find it so amazing that it haunted him like a recurring dream.[19] The keystone figures of the chapel windows are the prophet Mohammed, Confucius, Pope Julius II and Savonarola. The heads of Handel, Rossini, Schiller and Machiavelli are on the windows of the Art Gallery. Whilst eminent men are thus celebrated in remarkable diversity, the college's historian has observed that it is strange that no females are to be seen, such as, she suggests, Sappho, Hypatia, Boadicea or Queen Elizabeth I.[20] After the opening ceremony Martin-Holloway was knighted: Gleichen apologised profusely because he had not been able to arrange a baronetcy for him.

The governors, meeting in the Dean of Windsor's house, appointed as the first Lady Principal Matilda Bishop, Headmistress of the Oxford High School. She had studied at Queen's College under Maurice and had a near idolatry for the Anglican clergy. Miss Bishop introduced a daily service based on the Book of Common Prayer. Incensed at this, the Nonconformist press protested against the perversion of the Founder's design and at 'foul shapes of bigotry, priestcraft and chicanery stalking abroad which we thought had been banished for ever'. Eventually the Governors, upset by the publicity, gave instructions that Nonconformist services should alternate with those of the Church of England. Holloway's intentions regarding the membership of the Governing Body and the nondenominational character of the College had been disregarded. Now the Governors convened a conference in 1897 to discuss his educational policies. Almost the only defender of Holloway's plan that the College should award its own degrees was Strachan Davidson of Oxford, a prominent opponent of the admission of women to degrees there. The great majority of the participants recommended affiliation of the College to London University, which took place two years later.[21]

In 1912 the College was enabled to overrule Holloway's stipulation that women should not be eligible to be governors. Eventually in 1962 men were admitted as students; later the Royal Holloway College merged with Bedford College. What Thomas Holloway would have thought of these changes cannot be known. All that can be said is that as a business man he liked to move with the times and take the advice of experts; he could have accepted modernisation, but might well have been indignant at his favourite nephew's surrender to the Establishment and the Church of England.

HENRY SIDGWICK AT CAMBRIDGE

In 1903, shortly after Sidgwick's death, James Bryce wrote

> The England of our time has seen no movement more remarkable or more beneficial than that which has recognised the claims of women to the higher kind of education.... The change has come so quietly

> and unobtrusively that few people realise how great it is. No-one did more than Sidgwick to bring about this change.[22]

The campaign to enable women to take degrees at Oxford and Cambridge was a very long one. Whilst London University set the example in 1878, it took a further 42 years for Oxford and 70 years for Cambridge to follow it. In both universities the advocates of women's education tended to be divided into two groups, those for whom women's education was an end in itself, and those for whom it must provide a proof of their equality with men.

In Cambridge, Emily Davies (1830–1921) was the daughter of an Anglican clergyman. Though she had the necessary support of several influential Cambridge academics she was herself the moving spirit in persuading the Cambridge Local Examination Syndicate to open its examinations (equivalent to the later School Certificate or O-level examination) to girls as well as boys. These, unlike those of Oxford, were held all over the country. Equally important was her success in persuading the Royal Commission on Secondary Education of 1867–8 to include girls' schools in its terms of reference. The Commission's report concluded that 'there is weighty evidence to the effect that the essential capacity for learning is the same or nearly the same in both sexes, but that in middle class female education there is a want of thoroughness and foundation, slovenliness and showy superficiality'.[23] The report led to a rapid expansion of the kind of secondary education already provided by Queen's and Bedford Colleges and in the founding by Emily Davies of a college for higher education at Hitchin; this eventually moved to Cambridge and became Girton College.

Within Cambridge University itself the initiative was taken and the campaign masterminded for 30 years by Henry Sidgwick. Sidgwick (1838–1900) was the son of an Anglican clergyman who was Headmaster of Skipton Grammar School. At Rugby School he absorbed Dr Arnold's ethic of duty, which was never to leave him, even when he lost his Christian faith. He had a brilliant academic career at Trinity College, Cambridge, where, like Maurice, he was a prominent member of the Apostles. He was elected a Fellow of the College and Reader in the University; eventually he succeeded to Maurice's former Chair as Professor of Moral Philosophy. The variety of his academic interests can be seen from the titles of his three most important books, *The Methods of Ethics*, *The Principles of Political Economy* and *The Elements of Politics*. Over many years he was one of the most influential workers for the reform and modernisation of the University. He was only absent from

Cambridge for one term between his matriculation and his death. Like his close fellow Rugbeian friend, the poet Arthur Clough, he suffered a crisis of faith and at the age of 30 resigned his college fellowship because he could no longer accept the dogmatic obligations of the Apostles' Creed which were attached to it. 'For the rest of his life', Maynard Keynes was to remark, Sidgwick 'never did anything but wonder whether Christianity was true and prove that it wasn't and hope that it was'.[24] In taking up vigorously the question of higher education of women he was influenced both by Mill and by Maurice who was then residing in the University. But it also seems that he felt himself in search of a cause to which he could make a constructive contribution whilst wrestling with the religious doubts which had led to the negative gesture of resignation of his fellowship. He was not, like Mill, an ardent feminist, and only gradually came to support female suffrage. He saw the education of women as part of the process of the general reform of the University and hoped that experiments in a curriculum for women could help by their example to modernise the requirements for male undergraduates.

Like Maurice, Sidgwick was first drawn into practical action for education of women by sympathy for the needs of governesses and school mistresses; and he became a member of a board set up to examine them. But his interest now became broader. As he explained in a letter to the *Spectator* in 1869,

> the present exclusion of women from the higher studies of Cambridge University is perfectly indefensible in practice and must sooner or later give way. When the barrier is broken down, whatever special examinations for women may still be retained will be very different from what we now institute. Meanwhile two distinct classes must be provided for, intellectual girls who try to obtain honours and professional teachers who only need proof of capacity.[25]

Such a view faced stubborn opposition in influential quarters. *The Times*, about the same time, declared 'That education will always be the best for girls which is the most domestic, and English homes will always be the best schools in which English wives and mothers can best be trained.[26]

The first Cambridge Higher Local Examinations were held in 1869. Sidgwick now took the lead with Henry and Millicent Fawcett in organising lectures for women in the subjects of the new examination. The availability of lectures by some of the most eminent professors of the University and the prospect of examination by them drew girls from all over the country to Cambridge. Mill publicised and accelerated the attraction by endowing a resident scholarship, and other donors followed his example. Sidgwick, to his surprise and at his own expense, found himself renting a house to accommodate these students. He persuaded Anne Jemima Clough, the sister of his friend Arthur Clough and the Secretary of the North of England Council for the Education of Women, to come and run it. He wrote to a friend, 'I am going to have all the fun of being married without the burden of a wife.... I am growing fond of women. I like working with them. I begin to sympathise with the pleasures of the mild parson.'[27]

As the number of students attracted to Cambridge by the lectures increased, it was decided to build. A company was formed which raised money by subscriptions and shares and in 1875 Newnham Hall, later to become Newnham College, opened. Two-thirds of Cambridge's professors were persuaded to lecture to the women. With this demonstration of support, the examiners for the final tripos or BA were persuaded to examine the women informally at the same time. Eventually in 1881 the arrangement was formalised by the Senate and the names and ranking of the women were published, though they were not awarded degrees. For Sidgwick a very satisfactory element in this scheme was that the women could commence their studies after passing the Cambridge Higher Local Examination without taking Latin or Greek, whereas the men who took the University's preliminary examination had to pass in these languages. He considered the imposition of two dead languages on boys coming up to the University as having a very mischievous effect on education and hoped that it would shortly be removed so that the University would be opened to those from schools where Classics were not taught or who had studied on the modern side.

Meanwhile Emily Davies was proceeding on different lines. When she moved her college from Hitchin to the outskirts of Cambridge at Girton there would have been the possibility of a merger with the nascent Newnham. She insisted however that her students must take exactly the same courses as men, including the preliminary papers in Classics. She held this view so strongly that she described Sidgwick as 'a serpent gnawing at our vitals'.[28] Another difference was in religion. Emily Davies arranged that Girton, like the men's colleges of Cambridge, should be Anglican, whilst Newnham, under Sidgwick's agnostic influence, became nondenominational.

There was much to be said for Emily Davies's rigid insistence that women should have exactly the same education as men within the wider context of the campaign for women's rights and entry to the professions. Sidgwick's attitude on the other hand was always experimental.

He wanted women to have a *better* higher education than men. 'Many experiments are necessary before the exact form which the higher education of women ought to take is determined,' he wrote when he first became involved in the question. He experimented enthusiastically with correspondence courses, considering that they 'brought the student to make clear to herself the nature of her difficulties, which is more than half the way to solving them'.[29] He saw women's education as a social exercise. Education for him was above all a process of liberation which would increase happiness; women, he considered, ought to have the freedom to find out how they could best profit from what a university could offer. Sidgwick felt no hostility to Girton. He was a member of its Council and saw advantages in having two women's colleges in Cambridge, experimenting along different lines. What he appreciated from within the Citadel however was that there would be a very strong opposition to the University's cooperation in education of women if the question too soon arose of their participation in its government, wealth and privileges. He wanted women to have degrees but believed that Cambridge would accept filtration better than frontal assault. It was characteristic of his open mind that towards the end of his life he was prepared to admit the force of Emily Davies's position. 'We began with the notion (being inexperienced and not having worked it out)', he told the Conference on University Degrees for Women in 1897, 'of education adapted to women; and the whole process of our work has been towards realising that the one thing wanted was systematic study as it had been laid down by long experience for men.'[30]

In 1876 Sidgwick married Eleanor (Nora) Balfour, the niece of one future Prime Minister, Lord Salisbury, and sister of another, A. J. Balfour. The Balfours were feminists and keen amateur scientists. Nora was a capable mathematician who helped her brother-in-law, the brilliant physicist Lord Rayleigh, to correct examination papers. Although Sidgwick had been the tutor of her brother Arthur, and Nora had spent a term at Newnham, they only came to know each other well through their common interest in psychical research. They had no children, and Nora came to devote herself almost completely to Newnham. Shortly after her marriage she became Treasurer of the College. When the rapid expansion of its buildings and student numbers caused the work to become too much for Miss Clough, Nora became Vice Principal and the Sidgwicks consequently moved into the College temporarily for two years. On Miss Clough's death in 1892 Nora was her inevitable successor and the Sidgwicks again lived in the College, where Henry remained until his death. University friends admired his sacrifice of his privacy but he told his wife, 'the more I think of it the more I feel that the position of appendage to the Principal is one I was born to fit. . . . You will have all the responsibility for the entertainment and I shall have only the function of free critic.'[31]

For Mrs Sidgwick a conversation had to have a subject; she had no small talk. Henry Sidgwick on the other hand could talk indefinitely on both substance or trivia. One of his former students recollected how 'he would draw out the retiring, tolerate the absurd and welcome even the dull and commonplace. Our most fatuous remarks would be accepted and discussed.' Sidgwick's residence in Newnham gave it distinction and status at a time when the emancipation of women in both the educational and political sphere seemed to have come to a dead end. In 1887 Agnata Ramsay of Girton had headed the classical tripos list. In 1890 Philippa Fawcett of Newnham, the daughter of Henry and Millicent Fawcett, was even classed above the Senior Wrangler. There was now considerable pressure from all over the country for Cambridge to award degrees to women. Sidgwick had misgivings about the timing. He was well aware of the horror widely felt that if women were granted degrees they would come to have a share in the government of the University. Through his brother Arthur, who was a Fellow at Corpus Christi College, Oxford, he knew too of an alliance between conservative dons in Oxford and Cambridge which sought to divert the pressure for admission of women into a project for founding a separate university for women only. He feared that even the gains of 1881 might be lost. When the question came to the Senate however, in 1897, he pleaded earnestly that though the substance had been granted in 1881 the symbol was now also required when all the universities in Britain except Oxford and Cambridge had granted degrees to women. The outside world could not understand, he said, Cambridge's refusal, which appeared either unjust or stupid. But the proposal was refused by 1,701 to 661, largely by the votes of the MAs from outside Cambridge. Opponents egged on undergraduates to demonstrate. Some students feared that women graduates would bring more competition into a crowded job market. Others disliked the idea of female intrusion into the atmosphere of male camaraderie. An effigy of a gowned woman in bloomers was hung on a lamp post outside the voting place and the defeat was celebrated by destructive bonfires.[32]

The result of the debate and the manner in which it was conducted was hurtful to the Sidgwicks. Their college however continued to progress. In 1888, after the University ceremony at which three of the recipients of honorary degrees were Nora's brother, brother-in-law and uncle, the College gave a garden party to celebrate the opening of its third hall. This was attended not only by Lord Salisbury, A. J. Balfour and Lord Rayleigh but by the Archbishop of Canterbury, the Prince and Princess of Wales and three of their daughters. As the royal carriages drove up an avenue flanked by three hundred curtsying past and present students, all wearing irises, Sidgwick may have been well satisfied with what had been achieved. The increased happiness which Newnham had instituted into the lives of many of its students was, he said, the chief reward he had, or could have, for the time and money which he had spent on establishing it.[33]

Sidgwick died in 1900. The memorial to him in the College garden reads, 'The daughters of this house to those that shall come after commend the filial remembrance of Henry Sidgwick'. He is often described as the leader of the reform movement in Cambridge as well as that for higher education of women. As he saw himself however, 'there is nothing in me of prophet or apostle. The great vital joy giving qualities that I admire I cannot attain to.' He could not, he said, aspire to be a hero like the Positivists Frederic Harrison and E. S. Beesley, or Jowett at Oxford. 'Little people', he added sadly, 'should at least be harmless.'[34] Mrs Sidgwick found the 'florid tributes' which poured in on his death inappropriate for he and she, she said, were 'grey, grey people'. Oxford admitted women to degrees in 1920. When the question came before the Cambridge Senate a year later, once more the dinosaurs staggered in from their rural dens to vote. The proposal to give women full membership of the University was defeated by 904 to 712. Many of the clergy, who often had daughters at Cambridge or other universities, had been converted. Much of the opposition now came from the medical profession and the scientists who stressed the physiological and psychological differences between men and women. Once more male students rioted and even damaged the memorial gates of Newnham. Menaced by the possibility of interference by a Royal Commission and by Parliament, the Senate now conceded that women might write BA after their names. It was not until 1948 however that they became full members of the University with the same standing as men in its governance.[35]

HIGH CHURCH AND LIBERALS AT OXFORD

The development of women's education at Oxford was influenced by the conflict between conservatives and reformers and between High Church and Broad Church. A demand for systematic teaching of women was stimulated in the 1870s by the thirst for knowledge of the young wives of the first married Fellows and by the example of Ruskin as Professor of Fine Arts in inviting women to attend his lectures. In 1875 the University authorised special examinations for women. In 1878 an offer was received through the rationalist London preacher Moncure Conway to finance the establishment of a Hall of Residence and enable women from outside Oxford to profit from the courses now available.

Both conservatives and liberals reacted promptly to head off an initiative inspired by Conway. The High Church party was divided. Dr Pusey viewed the prospect of a women's hall as a great misfortune, and Dr Burgon in a famous sermon in New College told the women in the congregation, 'inferior to us God made you and inferior to the end of time you will remain'. Other High Church men however, notably Edward Talbot, the Warden of the experimental and successful Keble College, feared that if Oxford did not respond to the increasing demand for women teachers, with their immense potential influence, these would be left in the hands of Evangelical Cambridge and Godless London. The liberals would have been glad to cooperate but no agreement could be reached on the place of the Church of England in a Hall; so in 1878–9 Lady Margaret Hall was founded on an Anglican and Somerville on a nondenominational basis. These were eventually to become the first women's colleges. The Halls were linked by the Association for Higher Education of Women in Oxford whose secretary, Annie Rogers, organised the education programmes with such discretion that the issue of women's full admission to the University was never as bitterly disputed as at Cambridge.

Whilst Talbot was the most effective supporter of women's education among the High Church party, Henry Sidgwick's brother Arthur, Fellow of Corpus Christi College and Reader in Greek, was perhaps the most valuable ally among the liberals, mediating between those who wished to press for admission of women to degrees on the same terms as men and those who would be content with something less which exempted them from examination in Latin and Greek. Like Henry, Arthur early in life had renounced his Cambridge fellowship as a protest against the religious tests imposed. In the next generation another agnostic, Gilbert Murray, was prominent in the cause. Yet strangely it was Lord Curzon, a Conservative and President of the National League for Opposing Women's Suffrage, who as Chancellor of the University gave the final push which resulted in women being admitted as members in 1920, 28 years before victory was achieved at Cambridge.[36]

That Cambridge and Oxford granted degrees to women so much later than other British universities was due to their unique history and their curiously democratic form of government. On the one hand the traditions of centuries of clerical celibacy had to be overcome. On the other, all their MAs, whether working and living in these universities or not, had the right to vote in the Cambridge Senate and the Oxford Convocation, which had the final power of decision on policy questions. The Sidgwicks had done much to convert the resident teachers of Cambridge, except the scientists, to their views but the aging MAs outside, indignant at the prospect of change in the happy haunts of their youth, had to die out before equality for women was achieved.

If space permitted, many other men (though they were usually in a minority) who championed women's rights to equality in education deserve to be mentioned. Particularly influential were Lord Lyttleton and Arthur Hobhouse as members of the Endowed Schools Commission of 1868. Lyttleton described 'one of the grossest instances of injustice, one of the most unrighteous deprecations that can be mentioned, the withholding from the whole female sex of England for a very long time past of any benefit from the ancient educational endowments of the country'. His commission succeeded in opening entrance to many of the reformed grammar schools to girls.[37]

As the Principal of Somerville College, Margery Fry, stated in a tribute to Gilbert Murray in 1936, 'A certain period in the social history of England brought with it the need for a particular kind of disinterested and unpopular action, or particular form of co-operation between men and women demanding and creating confidence of no common order.' It was not charity which women had wanted, she continued, but a sense of being believed in, valued, wanted for the potential help in the life of the community, in the endless struggle against cruelty and needless suffering, which roused a peculiarly cordial gratitude in the women in the emancipation movement.[38]

7 Medicine

It is one of the lessons of the history of progress that when the time for a reform has come you cannot resist it. . . . Opponents, when the time has come, are not merely dragged at the chariot wheels of progress – they help to turn them.

James Stansfeld, article on 'Medical Women' in *Nineteenth Century* (July 1877)

The struggle for women to be enabled to become doctors was the earliest and most hard fought of their campaigns to enter the professions. The women claimed that what they were seeking was *re-entry* to the medical profession. The researches of Sophia Jex-Blake and others caused them to point out that from the Middle Ages to the eighteenth century women had been the primary healers in the community and had even been professors of medicine in the universities of Italy and Spain. However, the function had been gradually removed from the household duties of the wife and mother with the development of the profession outside the home and women had been excluded. By the nineteenth century they had been relegated to a subordinate role as untrained nurses and were even largely superseded as midwives.

By the early part of the nineteenth century the medical profession was divided into three classes; physicians, surgeons and apothecaries. Physicians had to hold an Oxford or Cambridge degree (which implied membership of the Church of England) and were thus considered to be gentlemen as well as members of a learned society. Surgeons were generally seen as skilled craftsmen, and apothecaries as tradesmen, though they both sought a higher status. Nevertheless, until 1858 there was no formal stratification and no legal bar to prevent anyone from practising medicine. Under the Medical Act of 1858 a Medical Council was set up to oversee education and publish a register of all qualified medical practitioners in Great Britain and Ireland. Nineteen universities or professional associations were designated as licensing bodies, whose examinations were accepted as awarding the qualification.

In order to obtain entry to the medical profession women needed three things, firstly training in a medical college, secondly the right to be examined by one of the nineteen licensing bodies, and thirdly the opportunity to obtain hospital experience before they could set themselves up in practice or obtain paid appointments. The women who made these demands almost all came from families of the upper middle classes. By the late 1860s, as has been seen, they were obtaining much better secondary education than had previously been available and even higher education, though without degrees. They were mainly motivated by a religious or humanitarian desire to serve other women and children and to escape from the boredom of a life which did not stretch their intelligence. They were supported by many women in the middle classes who disliked and resented receiving medical treatment from men.

A majority of medical men opposed their entry to the profession, though throughout the campaign they had allies within the Citadel. The doctors who opposed them argued that their study of medicine would lead to an indecent relationship between men and women, and that their mental and physical stamina was insufficient for its practice. Privately they feared that admission of women could jeopardise their own ambition for greater social recognition in a society in which gentlemen transacted the affairs of their professions in cosy all-male associations and clubs. Few of them admitted, but many believed, that the entry of women, who might capture the lucrative market in obstetrics and women's and children's diseases, would lead to unemployment of male doctors, including not only themselves and their sons, but their present students, to whom some of them expressed these forebodings.

Among laymen the reaction was more equally divided and opinion was much more influenced by the arguments put forward during the controversy than it was among the doctors with their vested interests. There was no clear division on party lines. The church gave no lead one way or the other. After a relatively brief campaign, Parliament enabled women to become doctors long before they obtained the suffrage. Its Members were largely influenced by the demands of women (including their own wives and daughters) to be able to consult women doctors; by the fact that the leading European countries including France, Germany and Austria had already taken action on the matter; and that in general the professions, widely accused of 'trade unionism', could not be trusted to reform themselves.

THE PIONEER DOCTORS

The obstacles encountered and the way in which they were overcome with the aid of a few committed men inside and ouside the profession can be seen in the careers of three pioneers, Elizabeth Blackwell, Elizabeth Garrett Anderson, and Sophia Jex-Blake. The earliest of these, Elizabeth Blackwell (1821–1910), was the daughter of a Bristol sugar manufacturer who migrated to America when she was 11. In New York and then in Cincinnati the Blackwells lived in a circle of abolitionists and reformers who advocated women's rights. When Samuel Blackwell died suddenly, his daughters kept the family going by running a school. Elizabeth had an unhappy love affair which left her with a longing to involve herself in some purpose greater than herself. A lady who was dying of cancer whom she visited asked why she did not devote herself to medicine, saying 'had I been treated by a lady doctor, my worst sufferings would have been spared me'.[1] With the possible exception of Harriet Hunt of Boston, who was not fully trained, no woman in America or Britain had at the time adopted medicine as a career. Doctors whom Elizabeth Blackwell consulted were discouraging. Her first need was to earn enough to finance her studies. She obtained a post as teacher of music in North Carolina in a school whose Principal was a doctor and allowed her to use his medical library. Now she experienced a religious revelation. 'A glorious presence, as of brilliant light, flooded my soul. All hesitation as to the rightfulness of my purpose left me and never in after life returned.'[2] After unsuccessful applications in Philadelphia and New York she applied in 1847 to a small college in Geneva, New York State. The Dean turned over to the all-male student body the decision about admitting her. They resolved

> That one of the radical principles of a Republican Government is the universal education of both sexes; that to every branch of scientific education the door should be opened equally to all; that the application of Elizabeth Blackwell to become a member of our class meets our entire approval; and in extending our unanimous invitation we pledge ourselves that no conduct of ours shall cause her to regret her attendance at this institution.

The innocent idealism of this statement shines out in a story in which later male students usually saw women as potential competitors, to be excluded from the profession by ridicule, by threats to quit any classes which admitted them and even by downright bullying. Elizabeth Blackwell was admitted to the class, who treated her like an elder sister.[3] She acquired considerable cunning. When during the vacation she requested permission to undertake practical studies in the Blockley Almshouse of Philadelphia, she lobbied the leaders of the three political parties –

Whigs, Democrats and Native Americans, each of whom was prepared to fight on her behalf against the expected obscurantism of the others. Whilst the senior doctor of the Almshouse gave her every facility, some of the younger ones obstructed her by removing case cards from the beds and ignoring her.

She returned to Geneva, passed all the final examinations with the best record in the class and was warmly cheered at the graduation ceremony. The press in Europe as well as America commented on her achievement in becoming the first qualified woman doctor. In London *Punch* urged

> Young ladies all of every clime
> Especially of Britain,
> Who wholly occupy your time
> In novels or in knitting,
> Whose skill is but to play
> Sing, dance or French, to clack well
> Reflect on the example pray
> Of excellent Miss Blackwell.[4]

Although she was now qualified, practical hospital experience was essential, particularly in women's diseases and midwifery, which were the obvious subjects in which to practise. Paris appeared to offer this experience. But she was refused either admission to the École de Médecine or to join the groups attending the chief physicians in the wards. Humbly she entered La Maternité, the foremost training centre in midwifery in Europe, as a nurse. The experience was invaluable, but whilst syringing the eye of a baby one of her own eyes was infected with purulent ophthalmia and had to be removed.

When she recovered she moved to England in 1850 where she was kindly received by the Dean, James Paget, and his wife at St Bartholomew's Hospital, London, where ironically, she was allowed on all the wards except in the department of midwifery and women's diseases. It was precisely in this department that male doctors were alive to the threat of female competition. Outside the hospital she was sought out by the group who were to become leaders of the movement for women's rights, including Barbara Leigh Smith (later Mme Bodichon) and Bessie Parkes, but she wrote home

> I cannot sympathise fully with an anti man movement. I have had too much kindness and just recognition from men to make such an attitude

other than painful. I think the true end of freedom may be gained better in another way... the object of education has nothing to do with men's rights or women's rights but with the development of the human soul and body.'[5]

Lacking the capital to set up in practice and with no family connections, she returned to New York. Whilst waiting to try to set up a practice there she gave a course of lectures on 'The Laws of Life with Special Reference to Physical Education for Girls', later published as a book. This won the particular approval of Quakers and when she opened a consulting room they provided the bulk of her patients. Eventually, with her sister Emily and Marie Zakrelewska, a Pole, both of whom had qualified at Cleveland, she opened a small New York Infirmary for Women and Children, financially supported through appeals from Horace Greeley, *The New York Times*, Lloyd Garrison's *Liberator*, and by Quakers. After the Infirmary was launched, Elizabeth visited England at the request of women friends who invited her to lecture there and stimulate interest in the entry of women to the medical field. Whilst she was in England she was entered as a physician on the new medical register of Great Britain, which could include the names of those who qualified abroad before, but not after, the Medical Act of 1858 was passed. Before she went back to New York she formed a small committee in London of ladies who undertook to interview and advise women who considered taking up a medical career. One member was Mrs Russell Gurney, whose husband's intervention in Parliament many years later was to be crucial in the admission of women to the profession.

Elizabeth Blackwell's Infirmary in New York grew into a medical school with herself as Professor of Hygiene. The pioneer stage was over in America. In Boston and Philadelphia as well as New York there were medical schools for women, and several men's colleges had accepted women students. She moved back to England where she practised and helped in Elizabeth Garrett's Dispensary. But she wrote back to Marie Zakrelewska in New York, 'you know I am not a natural doctor... my thoughts and active interest are chiefly given to some of those moral ends for which I took up the study of medicine'. Charles Kingsley, who told her 'you are one of my heroes',[6] involved her in Christian Socialism. She lectured widely on the prevention of disease through sanitation and hygiene and became Professor of Gynaecology at the London Medical School for Women. When ill health brought about retirement, she wrote a book, *Counsel to Parents on the Moral Education of their Children*, which was refused by twelve publishers because it urged that the facts of life should be explained to girls. She lived to be 89. Her achievement had been invaluable as a precedent for her successors in Britain and America.

If Elizabeth Blackwell was the pioneer of women in the medical profession, it was Elizabeth Garrett (later Elizabeth Garrett Anderson, 1836–1917) who did more than anyone to cause women doctors to be accepted as respectable in Britain. She started with the advantage of having a wealthy father, a self-made businessman in Aldeburgh, Suffolk. She was inspired to study medicine through attending lectures by Elizabeth Blackwell, in which the useless life of the 'lady' was contrasted with the role of the woman physician. Her father at first told her that her idea was disgusting but when he saw how many obstacles were put in her way he felt challenged to support her. A financial contribution from him to the Middlesex Hospital in London obtained her admission there to obtain experience and test her vocation, nominally as a surgical nurse. Informally she attended classes with the medical students and paid for private instruction from some of the staff, several of whom were willing to help her privately and singly but not publicly. On Sundays she attended F. D. Maurice's sermons. Over-confidence caused her to fail to disguise the superiority of her knowledge over that of most of the male students. After this had been conspicuously demonstrated at lectures and in patients' rounds, a number of medical students presented a memorial to the hospital saying that 'The promiscuous assemblage of the sexes in the same class was a dangerous innovation likely to lead to results of an unpleasant character', and that 'the presence of a female student at the Middlesex School had become a byword and reproach amongst similar institutions'.[7] Threatened with student withdrawals, the hospital committee excluded her from lectures, though she acquired some certificates of proficiency. The *Lancet* congratulated the students on their determination to get rid of her and prophesied that other schools would follow their example. She applied in vain to several other hospitals for permission to study. Each rejected her application on the grounds that no medical examining body would admit women candidates for degrees and that the schools would therefore be educating an illegal practitioner.[8]

The examining bodies of Oxford, Cambridge, Glasgow and Edinburgh Universities all refused her application. The Royal College of Surgeons of London, in rejecting her request to take its diploma in midwifery, stated that it would in no way countenance the entry of women into the medical profession. Finally she approached the Society of Apothecaries, which was licensed to examine 'all persons desirous of practising medicine'; after taking Counsel's opinion, it accepted her application. Acceptance was conditional upon her obtaining certificates of having followed courses in anatomy and medicine and sessions in clinical practice. She attended a course in physiology given by T. H. Huxley, a strong supporter of women's education, at which she met Sophia Jex-Blake.[9]

Guided by Emily Davies, she applied in 1862 to matriculate at London University whose Senate rejected her application by one vote. Her father now urged her to go to America where she would be sure of a degree, offering to pay the expenses. She wrote to him, however

> believing as I do that women physicians of the highest order would be a great boon to many suffering women, and that in order to have them the legal recognition must be given here, I think my work is tolerably clear and plain, viz to go on acting as a pioneer towards this end, even though by doing so I spend the best years of my life sowing that of which other students will reap. I feel very much that there is a divine and beautiful fitness in my being the one appointed to do the work. If I had a more decided genius for medicine or anything else it would be wasted.[10]

She now turned to Scotland where Sophia Jex-Blake provided introductions and her father accompanied her. Her application was refused at Edinburgh. In St Andrews it was at first accepted but the Senate then ordered the return of her fees. The students considered that the Senate had treated her unjustly and her father threatened litigation. The Advocate General of Scotland advised, however, that the Senate had no legal authority to admit her. The *British Medical Journal* was delighted, commenting, 'It is indeed high time that this preposterous attempt to establish a race of feminine doctors should be exploded.'[11] Russell Gurney, the Recorder of London and MP, however, wrote to her that lawyers had been much interested in the case and were mostly in favour of admission of women to universities. Her visit to Scotland was not wasted; Dr George Day, Regius Professor of Medicine at St Andrews, became her firm supporter, taught her and gave her certificates in anatomy and physiology. In Edinburgh the future Sir James Simpson, the pioneer of anaesthetics and Professor of Midwifery, who had previously befriended Elizabeth Blackwell, helped her to obtain clinical experience.

Her main problem now was to obtain teaching in anatomy. From Aberdeen University a letter came declining instruction on the grounds that ladies should not 'leave the spheres of usefulness which God had pointed out to them in order to force themselves into competition with the lower walks of the medical profession'.[12] Eventually she found a young surgeon at the London Hospital to work with her on dissection. One of his colleagues admitted her to lectures on comparative anatomy and she was able to obtain experience on the wards as a nurse. The older, more serious students helped her. Others, perhaps egged on by members of the staff, complained of the 'shabbiness' of a rival student pretending to be a nurse. She survived to obtain the necessary certificates. She also survived several proposals of marriage, including one from the blind MP and Professor of Political Economy, Henry Fawcett, who married her sister Millicent instead.

When at last she presented herself before the Society of Apothecaries, they refused to examine her. Newsome Garrett threatened a lawsuit and they capitulated, though they forthwith altered their regulations to prevent any other woman from following the same route. In 1865 she passed the examination as Licentiate of the Society of Apothecaries and was licensed to practise – 'I was in the fortress, as it were,' she wrote later, 'but alone and likely to be for a long time.'[13] She set up in private practise, taking only women and children as patients. Among them was Josephine Butler who said that she gained more from her than from any other doctor because she could tell her more than she could any man. She also opened a dispensary for the poor, which won considerable support from both inside and outside the profession.

Nevertheless the qualification of LSA seemed insufficient and in 1869 she applied to the Dean of the Medical Faculty of Paris University for admission in order to qualify for the MD. Two other foreign women applied at the same time. The University rejected her application, but the Dean then referred it to the Minister of Education. The Minister, knowing that Lord Lyons, the British Ambassador, who supported the application, enjoyed the confidence of the Emperor Napoleon III, prudently passed the decision to the Council of Ministers. It happened that the Emperor was ill and in his absence the Empress Eugenie, who presided, approved the application of the candidates on her own authority, expressing the hope that other young women would follow them. Elizabeth passed her public examinations in 1870 amidst general applause as the first woman MD of the Sorbonne. On her return to England she was elected a member of the new London School Board with the largest number of votes of any candidate. Shortly after this, at the age of 34, she married Skelton Anderson, a well-to-do shipping manager whom she had come to know on a hospital committee. Many of her friends were despondent, fearing her loss to the profession, but she resolutely continued to practise. Though they had three children she also still took a lively part in public life including a controversy with the eminent psychiatrist Henry Maudsley, who had written that higher education would lead to nervous and mental disorders in girls.

From its opening in 1877 she had an active role in the London School of Medicine for Women, of which she was Dean from 1883 to 1917. It was crucial for the success of the school that its students be admitted to the University of London's examinations. Their entry was fiercely resisted by Sir William Jenner, the Queen's Physician, who declared that 'he would rather follow his dear daughter to the grave than see her subjected to such questions as could not be omitted from a proper examination for a medical degree'; his daughter, later, became a militant Suffragette. When in 1878 the University decided to admit women to degrees, the Commissioners who drafted the new charter stated that they had been influenced 'by the moderation with which the views of women have been laid before us by Mrs Garrett Anderson'.[14] After she relinquished her position as Dean of the school, she retired to Aldeburgh in 1907 where she became England's first woman Mayor. No longer constrained in retirement to maintain the respectability of her profession, she marched in militant Suffragette processions and became a conspicuous figure in demonstrations throughout the country, including an attempt to 'rush' Parliament. Eventually, however, she felt unable to follow the extreme militancy of the Pankhursts. She lived to be over 80 but it was noticeable that she had offended the Establishment. No honour or decoration was ever offered to her by the Government. Amidst the earnestness of the woman pioneers Elizabeth Garrett Anderson retained a quiet sense of fun. 'The first thing you must learn,' she told the new students at the London Women's Medical School, 'is to behave like gentlemen.' Emily Davies, her sole woman colleague on the London School Board, sternly told her 'your jokes are many and reckless'.[15]

She was singularly fortunate first in the support of her father then of her husband. When they were engaged Skelton wrote to her, 'I mean to be a successful man of business, neither interfering with your pursuits, nor being interfered with by you, but having our confidences on all feasible subjects at off times of the day and week, mutually advising and fortifying one another.' He sat on her committees when she required a businessman. He guaranteed the finances of the London Women's Hospital. He ran her campaign for the London School Board. But he never interfered with her professional life. When they became engaged Elizabeth Garrett had written to her sister Millicent 'I am sure that the woman question will never be solved in any complete way so long as marriage is thought to be incompatible with freedom and with an independent career, and I think that we may be able to do something to discourage the notion.'[16] The example of their partnership was perhaps as important as any other contribution which Elizabeth Garrett Anderson made to the women's cause.

Elizabeth Garrett Anderson's approach was generally conciliatory, in the belief that her best contribution was to pursue an evidently normal medical career and family life and to convert by example. That of her contemporary Sophia Jex-Blake (1840–1912) was confrontational. Her father was a lawyer – a useful relationship since much of her time in the struggle had to be spent in legal research. He made her a handsome allowance but was shocked when she accepted a fee for teaching at Queen's College, 'accepting wages that belong to a class beneath you'. She was herself at first more interested in education than in medicine, teaching in Germany and making a tour of American schools and colleges, on which she wrote a book. Almost accidentally she found herself doing administrative work at the New England Hospital for Women in Boston and found herself 'getting desperately in love with medicine'. Her father paid for her to study medicine in New York at Elizabeth Blackwell's newly established Infirmary for Women, but on his sudden death in 1868 she felt obliged to return to England and look after her mother.[17] When other members of the family were able to take over this duty, Sophia sought to resume her medical studies. At London University the Registrar told her that its charter made it impossible for women to be examined for medical degrees. Oxford and Cambridge regulations appeared equally exclusive and neither offered a complete medical education.

Her thoughts turned to Scottish universities with their reputation for enlightened views in education and their freedom from ecclesiastical trammels. In 1869 she made her first application to Edinburgh University to attend lectures. A struggle commenced which lasted four years and, though unsuccessful, did much in the long run, through the publicity which it received, to bring about the admission of women to the medical profession. The constitution of Edinburgh University was cumbrous. The majority of the Medical Faculty and then the Senatus voted to admit her. The University Court, however, overruled the Senatus on the grounds that such an important initiative could not be taken to meet the needs of only one lady. The Court was influenced by an appeal from 200 medical students that women should be excluded from their classes on grounds of 'delicacy'. Sophia made two valuable friends in

Edinburgh outside the medical profession. David Masson, Professor of English Literature, was her supporter and adviser on tactics within the University. Alexander Russel, editor of the influential daily newspaper the *Scotsman*, gave the cause wide publicity as a consequence of which several other women joined her. The group now sought not admission to lectures but matriculation. They were accepted. All five passed the matriculation examination, four of them appearing in the first seven places. 'Regulations for Education of Women' were now published by the University. Under these, women were to receive instruction in separate classes from men, which professors were permitted but not required, to conduct.

An early incident indicated that the women's position was precarious. One of them, Edith Pechey, out of 231 candidates won the Hope Scholarship for Chemistry. The scholarship however was given to a male candidate who was lower on the list on grounds that Miss Pechey was ineligible because the women had studied in a separate class. Considerable publicity was given to the episode, *The Times* being severely critical of Edinburgh University for its unfairness. On the other hand many of the male students became hostile when they saw that women were capable of defeating them in competitive examinations. Even more ominous, strong opposition to the admission of women began to emerge within the medical faculty. This was led by Professor (later Sir) Robert Christison, who was Professor of Material Medicine and a Physician to the Queen and was over 70. He maintained that women did not have the intellectual ability and stamina of men and that if they were to enter the medical profession its standards would drop severely. He was the only person to have a seat on all four of the University's governing bodies. Influenced by Christison and by the students, many of whom paid their fees directly to the professors, some of the latter refused to conduct classes for women, who were thus obliged to obtain teaching in the extramural schools.

Within the medical faculty the most important friend of the women was Sir James Simpson, who died however in 1870. About this time Joseph (later Lord) Lister arrived as Professor of Surgery; although a Quaker he believed that any lowering of the barriers between sexes would lead to a decline in moral standards, and became an influential opponent of the admission of women. In the General Council of the University a motion calling for the relaxation of the restrictions which were forcing women students to obtain their tuition outside the University was strongly opposed by Christison who said that his stand was approved by the Queen. The motion was just defeated, by 47 votes to 46.

The only hospital where clinical experience could be obtained was the Edinburgh Infirmary; 500 male students signed a petition opposing the admission of the women, now known as 'The Edinburgh Seven', to the Infirmary and a widely reported 'students' riot' took place at which a mob of students, some drunk, attempted to block the entry of the women to the anatomy class in Surgeon's Hall and pelted them with mud. A committee for securing a medical education to the women of Edinburgh was however formed, chaired by the Lord Provost. Its members included Charles Darwin, Russell Gurney and Henry Fawcett. Christison's assistant Craig now brought an action for defamation against Sophia for accusing him of leading the riot while drunk. Though only a farthing damages was awarded, her costs amounted to £1,000 and were paid by the Lord Provost's Committee. There were further setbacks. The Presidents of the Royal Colleges of Physicians and of Surgeons of Scotland cancelled their usual engagement to attend the annual prize-giving ceremony of the extramural schools because women would be among the recipients. In 1872 a new board of managers of the Infirmary voted to admit women students to the wards but opponents blocked their entry by an appeal to the courts.[18]

As the time approached for the women to be examined and enabled to graduate, the University Court only offered certificates instead of degrees. The women took legal action and the Lord Ordinary of Scotland ruled that they were entitled to be examined for degrees. In a further victory the courts ruled that women students could be admitted to the Infirmary. There was however a serious, and indeed final, defeat when the University successfully appealed to Scotland's highest court, which reversed by 7 votes to 5 the Lord Ordinary's ruling that the women were entitled to examination for degrees, declaring that the University had acted illegally in allowing them to matriculate. Thus the first British university to admit women had closed its doors to them after four wasted years. To add to Sophia's humiliation, she had been so preoccupied with legal research and lobbying that she failed in her first professional examination.

What now was to be done? Elizabeth Garrett Anderson told Sophia that her want of judgement and temper had done great harm. In a letter to *The Times* in 1873 she said that the time was not ripe for medical education of women in Britain. They should instead obtain degrees abroad. Although these would not entitle them to registration, in practise, she said, colleagues would accept a woman with a good foreign degree. The example of women quietly making a reputation for themselves in the profession would overcome opposition and lead to a change in the law of registration. Sophia replied to *The Times* 'protesting as strongly as is in my power . . . we live under English law, and to the English law we must conform. . . . It is to the English Government we must look for a remedy.' Every woman who went to study abroad, she maintained, did incalculable damage to the cause of medical women in this country and helped to postpone the final and satisfactory conclusion of the whole question.[19]

Two things, she saw, were now necessary. An institution must be established which would provide a full medical training for women, and Parliament must be persuaded to change the law to enable women to be examined. She devoted herself with remarkable vigour and success to achieving both objects. A committee of which both Elizabeth Blackwell and Elizabeth Garrett Anderson, as well as T. H. Huxley and several prominent medical and scientific men were persuaded to become members, was set up to establish the London School of Medicine for Women. Funds were raised and premises found. The problem was to obtain instruction in a hospital. Here James Stansfeld's help proved invaluable. The Royal Free Hospital had at the time no teaching role and hence had no male medical students who could protest against the presence of women. It was also in financial difficulties. Stansfeld persuaded its governors to admit women students in return for very substantial fees, which he and others guaranteed. The school opened in 1874. Enrolment was successful and courses proceeded smoothly, but the students began to be worried as to whether they would be accepted for examination and by whom.

PARLIAMENT AND THE QUEEN

Meanwhile Sophia was conducting an intense campaign to bring pressure on Parliament. The unfair treatment of the Edinburgh Seven was widely attacked in the press. A perceptible feeling in favour of women doctors was building up in the country, reflected in petitions to Parliament. Among those who encouraged this was the well-known novelist Charles Reade. Very fully briefed by Sophia, he now wrote a novel, *A Woman Hater*, in which a central character is a 'doctress', whom the hero meets starving in London because her foreign degree does not allow her to practise. Through 50 pages she relates to him the whole story of the 'Edinburgh Seven', of whom she had been one before taking a degree in France. The hero, a wealthy squire, installs her as a doctor in his village, where his position is such that no one dares to question her qualifications and where she vigorously improves the health of the population by improving sanitation. At the end of the book Reade calls on his readers to challenge Parliament to have a full discussion on the question of women doctors 'so that no temporary pettifogging half measure may slip into a thin house, like a weasel into an empty barn, and to obstruct legislation of a durable principle'. Medical schools, he insisted, should be thrown open to all, male modesty being a purely imaginary article. Acceptance of women in the medical profession, he argued, would raise the self-respect generally of women, no longer compelled to waste three hours a day dressing and combing their hair, and devoting their energies to croquet, crotchet and coquetry.[20] *A Woman Hater* appeared as a serial in *Blackwood's Magazine* in London and in *Harper's* in the USA before being published as a novel in three volumes. Reade not only used the book to promote the cause but the cause to promote the book. 'Let us strike while the iron is hot,' he urged Blackwood, 'we shall serve a good cause and perhaps make a big hit.' He wrote to Gladstone, Selborne and other Liberal leaders hoping 'to serve a higher discussion in Parliament than if left to Temple Cowper and others of that ilk'.[21]

About this time Sophia explored one more avenue. It was found that possession of a licence in midwifery of the College of Surgeons involved the same medical curriculum as the MRCS and entitled its members to be placed on the medical register and to practise legally as doctors. Sophia and two other women applied to be examined. The College of Surgeons found that legally it had no authority to refuse to examine. The three examiners promptly resigned and no others were appointed. The College's arrogance strengthened feeling in and outside Parliament that the decision as to whether women should be admitted to the profession could not be left to that profession's members.

In 1874 Gladstone's Liberal Government was in office, though somewhat shakily. Stansfeld was in the Cabinet as President of the Local Government Board. Though very sympathetic, he told Sophia that he would not be the best advocate of opening medical examinations to women because the doctors hated him as a result of other controversies. He introduced her to the Home Secretary, Robert Lowe, who was MP for the University of London and had received a petition from one-third of its students urging that all the universities of Britain be empowered to admit women. Lowe would have tried to persuade his colleagues that legislation be introduced if they had not been preoccupied by the imminent resignation of the Government. Eventually the Liberal William Cowper Temple, who had carried through the Medical Bill of

1858 which had set up the examination procedures, took the initiative and introduced a Bill to enable the universities of Scotland to admit women as students and to grant them degrees. Though Edinburgh University's Senatus and Medical Faculty petitioned against it, the Edinburgh Town Council petitioned in favour. Before the debate Sophia lobbied indefatigably, armed with introductions from Stansfeld and Masson. When she could not obtain interviews with prominent politicians she approached them through their wives.

Introducing his Bill in June 1874, Cowper Temple reminded the House of Commons that it was its tradition to give a patient hearing to the grievances caused by defects in law when the victims were weak and had no voice in the election of Members. He pointed out that women could now take medical – and other – degrees in France, Italy, Austria, Germany and Switzerland and that 16,000 women had petitioned the House informing it of their desire to consult medical advisers of their own sex. The question could not, he maintained, be left to the doctors to determine, for the members of the profession were often unable to consider innovations relating to themselves without bias. In the subsequent debate the Conservative MP for Cambridge University, Beresford Hope, argued that women should remain nurses – declaring 'God sent women to be ministering angels, smoothing the pillow, whispering heavenly comfort.... Leave the physician's function, the iron wrist and iron will to men!' He protested against the perversion of women into deteriorated men. Stansfeld supported the Bill. Lyon Playfair, the MP for Edinburgh University, however, succeeded in having the vote postponed in order to give the University further time to consider the position.[22]

The Conservatives under Disraeli were in office when in March 1875 the debate was resumed and Cowper Temple introduced in the House of Commons a 'Universities (Scotland) Degrees to Women Bill' along the lines of his previous motion. He tried to allay the fears, in the mind of medical men, of competition and unemployment in the profession by arguing that women doctors would mainly be consulted by women patients who would not otherwise agree to be treated by men. There was also, he pointed out, a great demand in India for women doctors for the same reason. The Radicals supported the Bill warmly, Stansfeld and J. A. Roebuck saying that opposition to it was motivated by 'trade unionism' in the medical profession. There was a strong feeling however that the Bill sought by a side door to bring in higher education for women, a question which needed to be treated more broadly. It was defeated by 194 votes to 151.[23]

In July 1876 the persistent Cowper Temple returned to the charge, using a different approach, when he introduced an amendment to the Medical Act of 1858 to allow women who held degrees from universities in France or those of Vienna, Berlin, Leipzig and Zurich to be placed on the medical register. Of the 20 women then studying medicine in Paris, he pointed out, 12 were British. He recited the whole story of unfairness again, accusing the medical profession of silent, dogged, obstinate refusal, from feelings of prejudice, to allow women to enter the profession. The Bill was opposed however as unfair to men who might hold foreign degrees and as opening the door to recognition of low-grade degrees such as those alleged to be sold by certain American colleges. Stansfeld pressed the Government to state its views and intentions. Viscount Sandon, Vice President of the Council, explained that whilst the Government could not support the present proposal it was prepared to support one which would deal with the whole issue more broadly and that such a Bill was about to be sponsored by Russell Gurney. At this, Temple Cowper withdrew his motion.[24]

It was clear that as a Liberal he was now the wrong advocate of the cause. Russell Gurney, the Recorder of London, though a Conservative, was widely respected on both sides. He had been in charge of the Married Women's Property Act of 1870 and he and his wife had for long encouraged the pioneer doctors, Elizabeth Blackwell and Elizabeth Garrett Anderson. Gurney's Bill proposed that all the authorised medical examining bodies, and not just those in Scotland, be enabled to examine women.

Before determining its position the government sought the views of the Medical Council through the latter's President, Dr (the future Sir Henry) Acland, who was to play a vital part in bringing the profession to accept a change upon which Parliament was likely to insist. Unlike many of his colleagues, he had a broad education, studying Classics at Oxford before qualifying in medicine. A graduate of Christ Church, he was a friend of many of its influential alumni, including Gladstone, the future leader of the Liberal Party, whose admiration he first earned with a book on the archaeology of Troy, and Lord Salisbury, the future leader of the Conservatives, who shared his interest in scientific research. For many years Acland was Regius Professor of Medicine at Oxford where he looked after two of Queen Victoria's sons who studied there, and accompanied one of them, the Prince of Wales, on a tour of Canada and the USA. A devout Churchman and loyal member of the Establishment, he had greatly improved the status of science at Oxford by bringing about the creation of the University

Museum, which concentrated all the branches of science under its roof.[25]

In 1875 the Lord President of the Council wrote to Acland to ask the Medical Council's advice on the proposal being made in Cowper Temple's Bill to amend the Medical Act of 1858 to enable women who had taken degrees of foreign universities to be entered on the medical register. With Gurney's proposed Bill also in mind, the Lord President continued

> the Bill cannot fail to raise in Parliament the general question of whether women ought to be able to look to medical practice, or certain branches of it, as open to them equally with men as a profession and means of livelihood. The Government may have to express an opinion on the general question with regard to women who desire to obtain legal status as medical practitioners or on examinations or rules which prevent them from accomplishing this wish.

The Council's observations were requested on this wide issue also.

The Council was unanimous that it could not recognise the foreign degrees of women doctors, both because this would be unfair to male doctors whose foreign degrees were not recognised and because it had not the resources to assess the education given in foreign universities. Acland in his Presidential Address of 1875 made it clear that he shared this view. On the wider issue the Council was sharply divided and a committee was set up to consider the matter. Its report, which was not unanimous, was referred back. Acland in 1870 had suggested a compromise by proposing a special diploma and programme of training for women which would enable them to practise in the diseases of women and children or to be superintendents of medical institutions.[26] This, of course, had not satisfied Sophia Jex-Blake and her associates. In his Presidential Address to the Medical Council in 1876 he referred to the growing feeling that sex should not be a bar to admission to the medical register. This, he said, was part of a general movement in modern times, notably in England, America and Russia, which might cheerfully be accepted by some who would not spontaneously approve it. His remarks gave a clear if diplomatic lead.

The final text of the Council's report to the Lord President stated that it was 'of the opinion that the study and practice of medicine and surgery, instead of affording a field of exertion well fitted for women, do on the contrary present special difficulties which cannot be disregarded'. By a vote of 14 to 7, however, a crucial clause was inserted adding, 'but the Council is not prepared to say that women ought to be excluded from the profession'. The Council advised that the education and examination of female students should be conducted entirely apart from that of male students. Should the universities and corporations prove to be unwilling to admit women to degrees or as licentiates, the Council indicated that it could recognise other examinations if so authorised by Parliament. The minority was partially appeased by the approval of a covering letter to the resolution stating that in the opinion of many there were almost insuperable difficulties, moral as well as physical, to effective education and successful pursuit of medicine by women.[27] The Government, however, seized on the vital admission that the Council, despite all the difficulties, was not prepared to say that women should be excluded from the profession. Disraeli advised his followers not to oppose Russell Gurney's motion that the universities and corporations be enabled to examine women. Chastened by the increasing evidence of public opinion in favour of woman doctors, and weakened by the advice of the Medical Council, even the fiercest opponents abstained. The Bill was also unopposed in the Lords where it was introduced by the veteran Conservative philanthropist Lord Shaftesbury.

The problem now was to find one of the examining bodies which would make use of its powers. Edith Pechey, one of the Edinburgh Seven, persuaded the King's and Queen's College of Physicians in Dublin to do so. She was probably a more suitable advocate at this time than Sophia, who had written a letter to *The Times* insinuating that the examiners who had failed her at Edinburgh had been prejudiced, thus alienating T. H. Huxley and other friends of the cause. The Citadel was breached. Shortly afterwards, in 1878, London University opened all its examinations to women. The vote in Convocation on the new charter of the University showed that whilst a majority of the arts graduates were in favour, the majority of medical graduates were opposed.[28]

Sophia had by now qualified in Zurich and Dublin. The London School of Medicine, of which she had been acting secretary, had been her child. Stansfeld, as Chairman of its Council, pointed out however, when a permanent appointment was to be made, that the post required 'tact and judgement'. The qualities needed to direct a campaign were not felt to be those required to nurse an infant institution. Sophia was not appointed. Bitterly hurt, she returned to Edinburgh, where she established her own school of medicine for women, which collapsed after a student rebellion. She spent the rest of her active life in private practice in Edinburgh and set up a dispensary for women and children which successfully developed into a hospital.[29]

Women had found ways to be trained, examined and entered on the medical register. At first however, some of them did not find it easy to obtain sufficient private patients to support a practise, though there were places for them in the public service and almost unlimited opportunities to set up clinics for treatment of working-class women and children. The Queen's known opposition to women doctors helped to shape continuing hostile middle-class attitudes. Now Acland was able to do another service to the cause. Since Disraeli made her Empress of India, proposals which would benefit her Indian Empire had gained the Queen's keen interest and support. Acland through his excellent relations with the Royal Family arranged for the Queen to receive Dr (the future Dame) Mary Scharlieb, the first British woman to practise medicine in India, despite the attempt of the Queen's Physician Sir William Jenner to prevent this interview.

Mary Scharlieb, though herself English, was the wife of an Anglo-Indian (Eurasian) lawyer in Madras. Dismayed by the consequences of the refusal of both Hindu and Muslim upper-class women to consult doctors because they were males, she persuaded the Surgeon General and the Governor of Madras to allow her and three other women (probably Anglo-Indians) to enrol in the Madras Medical College whilst her sister came out to look after her children. After taking her licentiate there, she came to England to work for her MB. The surgeon to whom she had an introduction at Guy's Hospital disapproved of women doctors. Eventually however, she qualified at the new London Women's College and was about to return to India to practise when she was received by the Queen. As she learnt of the sufferings of the Indian women, the Queen exclaimed, 'How can they tell me that women doctors are not needed in India!' She gave Dr Scharlieb her photograph to show to her Indian patients, saying 'Tell them how glad I am that they should have medical women to help them.' She followed this up by asking the Vicereine, Lady Dufferin, to set up a fund to bring women doctors to India. No one again could quote, as Christison had done, the Queen as an opponent of women doctors.[30]

FRIENDS AND OPPONENTS

It is not easy to understand why in the middle and late nineteenth century some leading members of the medical profession supported the admission of women to it and why others were opposed. Although the opponents were in the majority a substantial minority was counted in all the votes on the subject. As Christison's sons and biographers commented sadly in their biography of him, the controversy was embittered by the fact that a portion of the garrison took part with the enemy.[31] Perhaps of all the doctors the support of Acland, though not the most overt, proved the most important. The position of Lord Lister, the recognised leader of the profession, is particularly puzzling. He was brought up a Quaker and remained one until at the age of 29 he was obliged to leave the Society upon marrying a Scottish Episcopalian. The Quakers were pioneers not only in education of women but in allowing them responsibilities. His boyhood friends had been Gurneys, nephews and nieces of the social reformer Elizabeth Fry. He had studied in the freethinking atmosphere of University College, London. In his innovative advocacy of antiseptics he tenaciously endured ridicule. He courageously refused Queen Victoria's personal appeal to oppose vivisection, telling her that the Royal Family's favourite pastime of shooting caused more suffering to animals than medical experiments. Yet on this one question he was a reactionary. He never allowed women to attend his lectures. He warned that if women were admitted as students they must also be allowed to hold hospital offices; the existence of young people under the same roof, however, intimately associated in consultation and in aid in emergencies, would, he predicted, lead to great inconvenience and scandal; and in any case the hospital duties, which often taxed men to the utmost, could not be given to women. He even opposed the employment of female attendants in India for examination and treatment of venereal disease and as late as 1897 maintained that the need for women doctors in India was not as great as was assumed.[32]

Lister's contemporary at Edinburgh Sir James Simpson (1811–70), whose introduction of chloroform made him almost equally famous, on the other hand was a firm supporter of the women doctors, especially in his own field of midwifery, making Emily Blackwell his assistant, teaching Elizabeth Garrett, and supporting Sophia Jex-Blake's application to study at Edinburgh. It may be significant that whilst Lister came from the middle class, Simpson was the seventh son of a baker in a Scottish village and had worked and won his own way through scholarships before being appointed to the midwifery Chair at Edinburgh at the age of 28. His introduction of chloroform in childbirth had been attacked as dangerous to morals and religion and he had delved deeply into theology to justify his position.[33]

The third eminent member of the Edinburgh faculty, Sir Robert Christison, was known as 'the Nestor of the University'. Christison described the proposal that Parliament enable universities to examine women in medicine as abominable – a violation of human rights. He declared that there was no evidence of an adequate demand among females to be educated in medicine nor of an adequate demand for their service; every ploughman's wife, he said, expects to be attended by a male obstetrician. In brief, he maintained that female medical practitioners, because of the female mind and frame, were with rare exceptions quite unsuited to the exigencies of medical and surgical practise.[34]

There were some in the middle who were sympathetic but did not care to rock the medical boat. Among these was Sir James Paget, Physician to the Queen, who had allowed Elizabeth Blackwell to attend his lectures at St Bartholomew's Hospital in London and later warmly approved of the initiative of the Government of Madras in providing medical training for women free of cost. When the question arose however of admitting women to the London International Medical Congress, of which he was President, in 1880, he told a colleague that whilst being 'rather lukewarmly in favour' he was influenced to take a negative position because of 'what I heard of some of the American and Zurich women doctors whom it would be difficult to exclude though few decent Englishmen would like to be associated with them'.[35] It was this conference that Queen Victoria, shortly before her conversion by Mary Scharlieb, refused to open if women doctors were participants.

Outside the medical profession, the aspiring women doctors found support not only from liberals and radicals but from lawyers who resented the growing pretensions of the medical profession.[36] Throughout the women's struggle in Edinburgh they received constant and invaluable support from two men there in particular. David Masson, Professor of Rhetoric and English Literature in the University, was the son of a stonecutter in Aberdeen and had studied for the ministry before becoming a journalist and editor of *Macmillan's Magazine*. He was Professor of English Literature at London University before coming to Edinburgh, where he became keenly interested in the higher education of women. He was supremely confident of victory. Lecturing in 1886 to the Edinburgh Ladies' Education Association, he declared that 'It is as if, all the world over, it had just been found out that men's minds are but half the minds in the world and the world were still rubbing its eyes at the amazing discovery.' Any opposition, he said, is 'an owl in the desert'. London University, he observed, had adopted policies which inevitably must lead to the admission of women and the Scottish universities sooner or later would follow. His main role was to fight the women's cause on university committees and to provide Sophia Jex-Blake with introductions into wider circles both in Edinburgh and in London.[37]

Whilst Masson was the women's champion inside the University it was Alexander Russel (1814–76), the editor of the *Scotsman*, who gave the cause such frequent and sympathetic publicity that opinion in Edinburgh outside the University came increasingly to support it. Russel was the son of a solicitor and was apprenticed to a printer before becoming a journalist. After editing several local papers, he was editor of the *Scotsman* from 1848 to 1876. Under him, it was said, 'the name of the paper became familiar in the obscurest cranny of the country, was able to influence greatly the wider political and public affairs of Scotland and introduce common sense in every social movement'. The unfairness and brutality with which the Edinburgh Seven were treated by the University and its students were described in the *Scotsman* with indignation, ridicule and sarcasm and its articles were reproduced in the English press. In the midst of the controversy Russel married one of the Seven, Helen Evans.[38]

An aristocrat who was firmly on the side of the women doctors was Lord Shaftesbury. Now near the end of a long life devoted to charitable causes, he was able to compare the obscurantism of the argument that women were not wanted in the medical profession with the opposition in his youth to railways and telegraphs. He not only introduced Gurney's enabling Bill in the Lords but used his influence to help considerably in the establishment of the London School of Medicine for Women.[39] William Cowper Temple (1811–88) was a particularly devoted aristocratic ally. He had been born into the very centre of the Whig Establishment. Son of the fifth Earl Cowper, he was nephew of one Prime Minister, Lord Melbourne, and stepson of another, Lord Palmerston. He held several posts in Palmerston's administrations and never feared to stir up vested interests whether in initiating the Thames Embankment or in opposing encroachment on commons. He was best known as author of the Cowper Temple amendment to the Education Bill of 1870, which banned denominational religious instruction in schools supported by the rates. Gladstone gave him no office in his administration of 1868–74; thus he was a free agent in taking up the cause of the women doctors. That he did this so doggedly was probably due to his feeling of personal responsibility because as the minister introducing the Medical Act of 1858 he had unintentionally allowed it to be so worded as to be used to prevent women from entering the profession.

Stansfeld's important support is consistent with his courageous championship of other women's causes, which have been discussed in

Chapter 3 above. His reflections on this campaign are of considerable interest. In an article on 'Medical Women' in the *Nineteenth Century* in 1877 he paid tribute to his principal political opponent, the Prime Minister Disraeli, 'for his real interest in the subject as a woman's question'. He noted that the women were 'immensely indebted not only to their supporters at Edinburgh University but to those whose embodiment of professional prejudice enabled the case to be put to Parliament not only as one of practical policy and right but of private and personal injury on the part of a public body'. 'It is one of the lessons of the history of progress', he continued, 'that when the time for a reform has come you cannot resist it, though if you make the attempt, what you do is to widen its character or precipitate its advent. Opponents, when the time has come, are not merely dragged at the chariot wheels of progress – they help to turn them.'[40] Ironically, he said, the violent, unreasonable opposition of Sir Robert Christison had done more to bring victory to the cause of the women doctors than all the efforts of the men who supported it.

GENERAL PRACTITIONERS VERSUS MIDWIVES

The story of the long opposition by the medical profession to the training and registration of midwives and its conclusion appears to confirm Stansfeld's observation. The French Government from Napoleon's time, as well as several German cities from the early nineteenth century, had financed schools for midwives. In Britain, about the same time, Evangelical societies had proposed the development of a profession of trained midwives in order to relieve female unemployment. British general practitioners however were strongly opposed to this suggestion, fearing that trained midwives would deprive them of fees. Their journal, the *Lancet*, in 1841 wrote that the women of England, happily for her sons, were wholly deficient in the moral and physical organisation necessary for midwifery. The founder of the Obstetrical Society went even further, declaring in 1847 'all midwives are a mistake. The office of midwife should be abolished and the word midwifery done away with on account of its derivation.' So the pregnant poor who could not afford doctors' fees were left to the care of the drunken untrained matrons such as Dickens' Sairey Gamp.

Florence Nightingale founded a school for midwives at King's College Hospital, London, in 1861 but it had to be closed because of puerperal fever. The feminist groups took little interest in midwives. There was one medical man however, Dr James Edmunds, a West End consultant, son of an Independent Minister, and a leader in the Temperance Movement, who took practical action. In advance of his time he realised that doctors were spreading in maternity wards infections which they had picked up in their medical and surgical cases and post-mortems. He sought to attract women from the professional classes who could be trained to carry out much of the work outside hospitals, and was supported in this by Acland who, as has been seen, favoured a diploma course for women in medicine. In 1864 Edmunds formed the Female Medical Society, which, though its committee included Lord Shaftesbury and Mrs Gladstone, was attacked by the *Medical Times* in 1872 as consisting of 'hopeless spinsters and sterile matrons'. Only a few women qualified at the Society's college, for those from the professional classes had higher ambitions than being midwives. When the London School of Medicine was opened to train women doctors Edmunds closed his college and transferred his energies to the Temperance Movement.

Stansfeld, whilst President of the Local Government Board from 1871 to 1874, proposed a scheme for registration of two classes of midwives, one educated and the other at a lower level. The *British Medical Journal*, however, in 1874 protested that literate midwives might read the letters of the ladies they attended and make out false death certificates. Those doctors who admitted a role for midwives mostly saw this as the work of home helps who would cook and clean – duties which would deter educated women from the profession. For years Dr Robert Rentoul successfully led the opposition to registration, openly declaring that dentistry and pharmacy had been given away and now there was a danger that midwives would rob the general practitioner of his income. Even Dr Elizabeth Garrett Anderson opposed registration as leading to competition for women doctors and a higher abortion rate.

Jesse Collings, the Liberal Unionist Parliamentary Under Secretary for Home Affairs, in 1902 described the General Medical Council in its attitude to the question as like the most overbearing trade union in the country. The alarmingly high level of medical rejections of army recruits in the Boer War led to a campaign for 'National Efficiency' which embraced the midwives' cause. An exasperated Government now set up a register for midwives, over which the General Medical Council and British Medical Association had no authority – they thus suffered a humiliating defeat at the hands of women's organisations, Parliament and Government.[41]

8 Religion

There is little doubt that the women of the Army have played a part in the general emancipation of women in the Western nations.... The women who marched at the head of the little bands of despised Salvationists were accustoming the public mind to the spectacle of women in command.

General Bramwell Booth (1926)

Whatever the position of women may have been in apostolic times – a somewhat uncertain question – they were admitted to no positions of authority by the Roman Catholic and Greek Orthodox Churches except as heads of convents and other female institutions. There were rare exceptions in early mediaeval England when Abbesses of Anglo-Saxon royal families ruled over double monasteries of men and women. The Normans however were a warlike people who tended to allow a lower place to women than the Anglo-Saxons; under them even female convents were seldom ruled by an independent woman but usually by a Prioress who was responsible to an Abbot.

After the Reformation women became prominent in some of the sects which separated from the Church of England. Thus it was a group of women in Bedford who took the initiative in setting up the church which John Bunyan joined. The prophetess Elizabeth Poole was given an extensive hearing in the Whitehall debates of Cromwell's armies. It was among the Quakers alone however that the ministry of women continued to be recognised after that time, except for a brief period following John Wesley's death, when there were Methodist woman preachers. Other Nonconformist women ministers only began to be regularly appointed in the early twentieth century. In 1926 there were seven Unitarian women ministers, two Congregationalists and two Baptists. Anglican women priests were not appointed until the end of the century.

In this chapter the early contribution of the Quakers and Unitarians to recognition of women's spiritual equality is first considered. Next came the implications, startling at the time, of General Booth's decision to give women equal rank with men in the Salvation Army. Finally, because the Church of England's position as the Established Church was of particular importance, the way in which the ordination of women came about, after the surprising actions of two Bishops of Hong Kong, is discussed, even though this occurred after the period covered generally by this study.

QUAKERS AND UNITARIANS

From its beginnings in the seventeenth century the Society of Friends (Quakers) recognised the ministry of women on an equal footing to that of men. Its 'Nursing Mother' Margaret Fell, who was eventually to marry the founder George Fox, wrote a tract, *Women's Speaking Justified*, and visited Charles I and James II to persuade them to release Quakers from prison. A number of Quaker women went abroad as itinerant missionaries in the seventeenth and eighteenth centuries. In the nineteenth century they could be appointed as ministers and elders by their local meetings; in their capacities as ministers they were able to travel and preach at home and abroad, often at the expense of the Society. It was thus as a minister that Elizabeth Fry was able to conduct her campaign for prison reform. In the early nineteenth century women ministers outnumbered men.

Yet whilst women were treated as equals in spiritual matters they were not so on policy and business questions. Until the end of the nineteenth century the business of the Society was transacted in yearly, quarterly and monthly meetings, at the national, regional and local levels, in which men and women generally met separately, though there were some exceptions at the local level.

Although the experience of the Quaker women in running their own affairs gave them self-confidence and experience which was to be of considerable importance in the feminist movement, the opinions of the women's meetings had little influence on the Society's decisions, and until 1896 women could not be members of Meeting for Sufferings, its national executive committee. Whilst Meeting for Sufferings was pressed by some of its members early in the twentieth century to make a declaration about the position of women, when it eventually did so in 1914 it failed to commit the Society to support of women's suffrage.

No prominent Quaker man made the improvement of the position of women his main interest, as did the Unitarian James Stansfeld, with the exception of Jacob Bright whose role in Parliament is described in Chapter 10. This is partly because Quaker women were experienced in speaking in public and did not need men as mouthpieces. Thus they were the earliest collaborators with Josephine Butler in her campaign against the Contagious Diseases Acts though later some Quaker men also became active supporters inside and outside Parliament in this particular cause. Most of these were motivated in this as much, if not more, by dislike of extension of the power of the State as by a regard for women's rights. It was, however, a young Quaker publisher, Alfred Dyer, who developed a concern about child prostitution and whose vivid description of the Belgian brothels into which young British girls were lured triggered Josephine Butler's subsequent campaign to change the law on the Age of Consent. In general in England the Quakers' interest in the improvement of the position of women was nothing like as strong as that shown by them in the questions of slavery, temperance or prison reform. Further, some eminent Friends had a conservative and patriarchal attitude, notably John Bright, the best known Quaker in the nineteenth century, who declared that Mill's *The Subjection of Women* was pernicious, and spoke strongly against women's suffrage in Parliament. He maintained that women's place was in the home and that if they engaged in politics they would undermine the institution of the family, neglect their domestic duties and increase the influence of priests. As for those of them who asked for rights, he commented, 'My gardener says there is nothing he so dislikes in poultry as a crowing hen.' His brother Jacob and his daughters on the other hand were deeply committed to the women's cause. The Quaker numbers were steadily declining until 1860 mainly because those who married outside the Society were expelled from it. They were somewhat eclipsed by the Unitarians in the mid-nineteenth-century movement to demand improvements in the status of women. It is significant that Jacob Bright, though still a Quaker when he entered Parliament, later became a Unitarian.[1]

The Unitarians exercised an influence on the early stages of the feminist movement quite disproportionate to their number, which was only about 50,000 in Great Britain in 1851. By the nineteenth century they had abandoned the Calvinistic doctrine of predestination held by the Presbyterians. Liberated from the burden of original sin, and believing that Christ was not God but a perfect man, they had an optimistic belief in perfectibility – of women as well as of men. They derived from John Locke the belief that God had given the power of reason to both men and women, whose duty was to use it in order to discover the laws of the universe and then follow them. From their most important eighteenth-century leader, Joseph Priestley, they learnt that the study of science would bring about the mastery of nature, the increased well being of mankind, and show forth the glories of God.

Their approach to the Bible was critical. Their rational ethics differed sharply from the emotional fervour of the contemporary Evangelicals who emphasised the submission of women in obedience to biblical texts. Even more than the Quakers they had a passion for education. Unable to enter the Public Schools or to graduate at Oxford and Cambridge, they ran their own schools and academies. Priestley maintained that women had the same moral duties, dispositions and passions as men and that the education of girls deserved as much attention as that of boys. Girls were admitted to many of the Unitarian schools, whose curriculum, which favoured science and languages over the Classics, was helpful to them, though they were unable to follow their brothers who went on to the Scottish universities or to Leiden.

The education of their schools was reinforced by that in the home. Thus Octavia Hill (1838–1912) was taught by a Unitarian mother who had written articles on education and started the first industry for middle-class women, as well as by a grandfather, Southwood Smith, the great sanitary reformer, who had once been a Unitarian minister. Beatrice Webb grew up in a family of sisters with whom their father, a wealthy Unitarian industrialist, used to discuss the important decisions in his business at the breakfast table. William Smith, the Unitarian Parliamentary reformer, had three remarkable granddaughters. One was Florence Nightingale. The others were the early feminist leaders Bessie Parkes and Barbara Leigh Smith (later Mme Bodichon). What, Bessie asked Barbara, were Unitarian women to educate themselves *for* – 'I feel possessed of all sorts of faculties and constrained to put them to a use.' The middle class, she bitterly observed, threw numbers of mature women into the drawing room so they could mutually devour each other's time. Women from Unitarian families were to solve this problem for themselves by creating new professions. In addition to Octavia Hill in housing, there was Mary Carpenter in industrial schools and Barbara Leigh Smith and Bessie Parkes setting up an agency for women's employment and launching the first feminist journal; whilst Florence Nightingale who established the nursing profession, though an Anglican, was influenced by her Unitarian descent. [2]

Thanks to their education in science and their openness to innovation, Unitarians prospered as manufacturers in the industrial revolution. Their enthusiasm for laissez faire, for equality of opportunity and for political reform was in the spirit of an age dominated politically by the Liberal Party. They were the largest group of dissenters in the House of Commons. Yet with prosperity came the desire to conform to upper middle-class attitudes. Unitarianism, though legal since 1813, was still not quite respectable with the gentry; thus it was said that hardly a Unitarian remained one, once the family started to keep a carriage.

At the end of the nineteenth century, when Social Darwinism and Imperialism became fashionable and when prim socially conservative attitudes spread in the middle classes, the early élan of the Unitarians seemed to have died down though some, such as Pethick Lawrence, were active in the womens' suffrage movement. Their main contribution to the equality of women had by then been made. Unitarians had taken the lead in founding the universities of London, Manchester and Birmingham, firstly with no discrimination in religion and later with none in sex. They had played an important part in establishing women's colleges at Oxford and Cambridge. In general they had insisted, as one of their ministers, H. W. Crossley said, that maintenance of absolute freedom of thought on all conceivable questions was the chief aim of life. Within this philosophy the infant cause of womens rights had been able to grow.[3]

If any male Unitarian were to be singled out for his practical contribution it would be James Stansfeld who, as seen in other chapters, whether in office or from the back benches in Parliament, championed almost every women's cause over a period of forty years. But the man who perhaps did more than any contemporary except J. S. Mill to further the ideas of feminism was W. J. Fox, the minister of South Place Chapel, London. Indeed he exercised an important influence on Mill through Harriet Taylor, who attended his chapel and wrote under his editorship. As well as Mill, among the writers and changers of social attitudes who listened to his sermons were Carlyle, Dickens, Bulwer Lytton, Browning, George Eliot, G. H. Lewes and Leigh Hunt. To them he preached that literature should have a social and political function and should challenge the fashionable portrayal of women as perfect angelic beings. He strongly rebuked the Chartists, with whom he was otherwise sympathetic, when they excluded women from their demand for adult suffrage.

William Johnson Fox (1786–1864) was prominent not only as a minister of religion but as an editor, journalist and politician. Carlyle said that his eloquence was like the opening of a window through a London fog. The son of a weaver, he was largely self-educated except for taking a course at the Homerton College for Independent Ministers. South Place Chapel was built specially for him and at its opening women sat down together with men for the first time at a public dinner in London. His journal *The Monthly Repository* gave far more attention to women's rights than any of its contemporaries and published the earliest articles of Harriet Taylor and Harriet Martineau. In it he declared that the legal position of women in Britain was as inferior as that of slaves in

America. He maintained that physical weakness was the only natural deficiency of women and saw female emancipation as part of a wider process by which society might come to be ruled by reason, not force. 'Savage man', he wrote in 1833, 'kills and beats woman and makes her toil in the fields; semi civilised man locks her up in a harem; and man three quarters civilised, which is as far as we have got, educates her for pleasure and dependency, keeps her in a state of pupilage, closes most of the avenues of self support and cheats her by the false forms of an irrevocable contract into a life of submission to his will.' Later, in 1856, he strongly attacked the laws by which married women had no right to property. The capabilities of women were unfolding, he said, as their sphere of occupation was enlarged and women were increasingly becoming breadwinners. He gave instances of women singers who were married by adventurers in order that they might live off their voices. He suggested sardonically that the Anglican wedding service should not have the bridegroom say 'with all my worldly goods I thee endow' but 'with all thy worldly goods thou me endowest, and with what thou may earn or inherit in future'.[4]

He would pick on all sorts of things as pegs upon which to hang articles to illuminate the unfair lot of women. Reviewing a complacent Methodist's book on the Wesleys, he pointed out that though John, Charles and Samuel had brilliant and satisfying lives, their sister Mahetabel, a sensitive poet, was compelled by the family to marry a boorish, semi-literate plumber: when he treated her without affection they assured her that to endure misery was her duty.[5] It was when he began to argue that marriage should be made a civil contract and divorce made easy that Fox lost much of his congregation; for about that time he had parted from his wife and, though the relationship was probably paternal, a beautiful young ward lived in his house without a chaperone. He separated from the Unitarians and after running South Place Chapel as an independent minister for a time, he ceased to call himself Reverend and devoted himself to politics. He played a leading part in the campaign for the reform of the Corn Laws and was a Radical Member of Parliament from 1847 to 1862. Among his favourite causes were compulsory education and extension of the franchise. Fox, unlike Stansfeld, did not himself bring about reforms which benefited women, for he lived at an earlier time when Parliament was less aware of their problems. Indeed as a reformer he was generally some twenty years ahead of his time. His sermons, articles, speeches and personal influence were more important than his direct achievements. He deserves perhaps to be better remembered by historians of the feminist movement.

A striking characteristic of the mid-nineteenth-century Unitarians is the variety of their feminist activities. The contribution of William Shaen (1820–87) paralleled that of Stansfeld, but outside Parliament. After studying at UCL and Edinburgh University he became a barrister and then a solicitor and was convinced that the law ought to be equal between men and women. He could not bear, says his biographer, to see the advantages of higher education limited by the accident of sex: he became Chairman of the Council of Bedford College, London, and played an important part in the founding of Girton and Newnham at Cambridge and Somerville at Oxford. He incurred ridicule by proposing in the Senate that London University should admit women, 25 years before this came about. He was an early advocate of medical education for women, chairing Elizabeth Blackwell's lectures on the subject in 1858 and later helping to launch the London School of Medicine for Women. He gave free legal advice to bodies advocating equal rights for women in divorce and in suffrage, the repeal of the Contagious Diseases Acts and better wages for needlewomen. He acted as manager of Octavia Hill's first slum housing project. A Radical and Republican, he looked up to Mazzini as his master, who taught him, he said, 'the Religion of Progress, the harmonious organisation of the whole human race through the ever enlarging series of the divinely appointed spheres from the family, through the Commune, the City, the Province, the Nation and at length the concourse of all peoples'.[6]

WILLIAM AND BRAMWELL BOOTH

The women who led the campaigns for the suffrage and for entry into the professions, as well as the men who helped them, nearly all came from the upper and middle classes. So did Josephine Butler and her allies. The Salvation Army on the other hand drew its support mainly from the lower middle and working class. An unusual characteristic was that the Army was not organised on democratic lines but under the dictatorship of General William Booth (1829–1912). What appeared unique, however, even startling, to nineteenth-century contemporaries was that women in its service not only preached, conducted marriages and funerals but, most remarkable of all, often had command over men. As Bramwell Booth, the founder's son and successor, wrote in 1925,

> though seldom acknowledged, there is little doubt that the women of the Army have played a part in the general emancipation of woman

which we see to be going on in the western nations. . . . The women who marched at the head of the little bands of despised Salvationists in years gone by were accustoming the public man to the spectacle of woman in command – of woman taking an unshrinking share in public duty, and overcoming by the grace of God her supposed inferiorities.[7]

The initiative for this equal status of women came from the founder's wife, Catherine (1829–90), the 'mother of the Army'. As a girl Catherine Mumford, often bedridden, had much more time and inclination for reading, particularly of theology, than her future husband, who was working as a pawnbroker's assistant when they met. Her reading had convinced her that there was no scriptural justification for the subordinate position of women. The correspondence between William and Catherine during their courtship was crucial in establishing the balance of their own relationship and eventually the equality of women with men in the Salvation Army. William quoted in a letter an aphorism that 'woman has a fibre more in her heart and a cell less in her brain'. Catherine responded indignantly, telling him that woman was man's equal and that any inferiority was due to lack of training and opportunities. She would never, she said, take as her partner in life one who was not prepared to give woman her proper due. 'The heaving of society in America (the birthplace of so much that is great and noble),' she told her fiancé, 'though throwing up, as all such movements do, much that is absurd and extravagant and which I no more approve than you, yet shows that principles are working and enquiries awakening. . . . May the Lord . . . overrule the swaddling-bands of prejudice, ignorance and custom which almost all the world over have so long debased and wronged her.'[8]

William was by nature an autocrat but he listened to Catherine as he did to no one else. It was she who was to convince him to break with the Methodists and become a full-time evangelist. A curious incident brought about his complete acceptance of Catherine's views on the equality of women. Whilst William was still with the Methodists and stationed in Gateshead, a visiting American evangelist in nearby Newcastle was assisted by his eloquent wife. A local Nonconformist minister launched an attack on them, asserting that there was no scriptural authority for women to preach. Catherine was provoked by this to write a 30-page pamphlet, *Whether the Church Will Allow Women to Speak in Her Assemblies*. This required considerable research and in order to reveal any possible weaknesses in her argument she discussed the whole issue with William. In these long conversations about the scriptural evidence he came fully to accept her view and her conclusion that 'whether the Church will allow women to speak in her assemblies can only be a question of time'. Shortly after this in 1860, whilst in the minister's pew, Catherine suddenly felt the call to speak. Her husband, taken by surprise, invited her to do so. Her testimony so moved the congregation that she was asked to speak regularly and a little later replaced William in the pulpit when he fell ill. [9]

When William broke free to run an independent evangelistic mission which eventually became the Salvation Army, Catherine preached to the middle class in the West End whilst William spoke to the masses of the East End. She was a moving preacher and the financial contribution which her work brought in at this time was larger than that of William. It was Catherine who designed the Army's flag and uniforms and gave teetotalism a prominent place in its programme.

Despite Catherine's example, it was a lack of resources as much as principle which caused Booth to begin to use women as preachers in his independent ministry. It was found that women, even when reluctant and hesitant, could often win sympathy in open air meetings in the East End of London better than men. Before he founded the Army Booth's independent Christian mission was, however, organised by committees. His early lieutenant George Railton wrote in 1887

> I am not sure that we have ever had an instance of woman's successful management in association with a committee of men. A strong willed man – any man in fact with sufficient ability and strength of mind to be a successful officer – might impose his will upon a committee without offending them or driving them away; but no woman could do this without conflict that could destroy her influence and make her usefulness impossible on any extended scale. They never could have been sent out with any prospect of success had they been saddled with the old machinery.

It was only when Booth in 1878–9 transformed the mission into a dictatorship – the Salvation Army – Railton continued, that the women, untrammelled by committees, were able to exercise their full potential as leaders, and a cloud of women was suddenly sent flying over the country.[10]

William Booth firmly asserted the principle of equality in the Army's regulations. In the first orders and regulations for field officers in 1886 the Preface stated, 'In the Army men and women are alike eligible for all ranks, authorities and duties, all positions being open to all ranks.' In the 1900 edition this was elaborated – 'One of the leading principles upon which the Army is based is the right of woman to an equal share with man in the great work of publishing salvation to the world. By an unalterable provision of our Foundation Deed she can hold any position of authority or power in the Army from that of local officer to that of the General.' 'Find out the powers of our women candidates,' he urged, 'and give them the chance to use them for the glory of God.' He defended the activities of his Hallelujah lasses with the example of Miriam in the Bible, saying that it was absurd that women could legally sell, sing and dance in public but not prophesy. He proudly asserted that the example of the Salvation Army had helped to open positions of remunerative employment and trust to women outside it.[11]

Yet even in the early Salvation Army there was some opposition to the idea of equal status between the sexes. As Bramwell Booth recalled, men who had been raised up from the very gutters objected to women being regarded as on an equal level to themselves. Some married women, he said, thought it unwise and even indelicate to bring their single sisters into any confidential relations with male members or indeed into any public service in which they had to deal with men. Some husbands too, objected to their wives undertaking any sort of public work. The women who joined the Army were often better educated than the men. Even Catherine, who was recognised from the beginning as the Mother of the Army, though she took equal status for granted, at first hesitated at placing women in positions which involved authority over men as heads of stations, but agreed under pressure from Railton. At the senior level captains who had been quite happy to work side by side with women of the same rank objected, sometimes strongly, to their promotion above them to be Divisional Commanders, a position which was considered equivalent to that of an Anglican or Roman Catholic bishop.

One reason why Booth had separated from the Methodists was that they appeared more interested in evangelisation abroad than at home. When he parted from them he preached and recruited in the poorest quarters of the poorest towns of England, reaching people who would never attend a church or chapel. Catherine, who started the practice of visiting the homes of the very poor at night, proved that women officers had an essential role in reaching out to them. Booth had set out purely as an evangelist. The conditions which he found, however, caused the Army also to become a major relief agency, setting up feeding centres, shelters for the destitute and rescue homes for girls. Much of this work could only be undertaken by women.

William Booth had little interest in theology. At an early stage the Salvation Army gave up the use of sacraments; thus in practising equality between its male and female officers it never became involved in the question which came to torment the churches, as to whether women were qualified to celebrate Communion. Booth was prepared for the Army to run side by side with the Church of England 'like banks of a river with bridges thrown across over which the members could mutually pass and repass'. When discussions took place between Bramwell Booth, the Army Chief of Staff, and Randall Davidson, the future Archbishop of Canterbury, on behalf of the Church of England, the status of the Army's women, however, emerged as a major obstacle to cooperation. The equality of women was firmly established as one of the Army's basic principles. Davidson on the other hand could only suggest that consideration might be given by the Church to the appointment of some Army officers as deaconesses.[12]

It was from the Booths' own family that most of the women came whose careers conspicuously illustrated the principle of sex equality. At their mother's knee, the children learnt the lesson that boys were not superior to girls. She presided at table even when William was present. There were eight of them, Bramwell, Ballington, Catherine (Kate), Emma, Herbert, Marian, Evangeline (Eva) and Lucy. All except the invalid Marian were dedicated by their parents to the Army and held high positions in its service from a very early age. Of the girls, Emma, known as 'The Consul', was appointed to direct the Women's Training Home at the age of 20. Eva headed a large London Corps before commanding in Canada, and later the USA. Lucy at the age of 24 commanded the Army in India. The most spectacular early career was that of Kate, who, in 1881 at the age of 22, was sent to 'raid' France with three other girls and no male companions. Exceptionally, a woman who was not a Booth, Hannah Ochterlony, was appointed to head the Army in Sweden where the campaign had a rapid success.

When the daughters married, their husbands added Booth to their names and ruled jointly with them. Kate married Colonel Clibborn, a colleague who had come to sustain her in her Swiss campaign. The Booth-Clibborns went on from the command of France and Switzerland to that of Belgium and Holland. Emma married Frederick Tucker, the Army's Commissioner in India. When the Indian climate disagreed with her, they were transferred to the USA where she died in a rail accident. Lucy had the misfortune to be jilted by a fellow officer who broke off their engagement; an Army commission of enquiry decided that he must be insane and she was sent to take over the command in India where she married a Swedish Salvationist. As Lucy Booth-Hellberg she rose to the top rank, under the General and Chief of Staff, of Commissioner.

When William Booth died in 1912 he was succeeded by his eldest son Bramwell. At this time women officers considerably outnumbered men and Bramwell constantly sought to identify and promote those who were outstanding. He was sensitive to the loneliness of unmarried women officers in senior posts and tried to arrange for them to live with other women officers as companions. When women Salvationists married he requested his Commissioners to see that they continued to have an active role. He deplored the tendency to relegate women to posts where their services came cheaper because women's allowances were less than those of married men. Yet even in the Army there were men who had only tolerated the equality of women because they had been ordered to do so by the Founder. The Salvation Army archives are full of instructions from Bramwell to his reluctant senior officers urging the need to promote women.

Bramwell's championship of women may indeed have been a factor in his deposition by the Army Council when he was old and ill. It was not known whom, in accordance with the trust deed prepared by the Founder, the General would designate as his successor, in the sealed envelope to be opened after his death. It was supposed by certain senior officers that Bramwell would appoint his daughter Catherine; some did not care for her and others disliked the prospect of female rule. As it turned out, Bramwell's intention never became known. The sealed envelopc containing the name was formally burned unopened after he was removed from office. The Army High Council elected the Chief of Staff, Edward Higgins, as the new General. When Higgins retired five years later Evangeline Booth, at the age of 69, was elected General, though only on the fifth ballot; the principle of the equality of women, at least if they were members of the Founder's family, was thus reaffirmed. The wheel seemed to turn full circle. A hundred years after Railton had explained how under Booth's dictatorship women had achieved an equality which had been denied them under the previous committee system, articles by women officers in *The Officer* and *War Cry* complained that now that the Army was governed by Conferences and Boards women had been replaced in senior posts traditionally held by them in Britain and that in the USA there was no woman among its top 24 leaders.[13]

TWO BISHOPS IN HONG KONG

Within the Church of England sisterhoods were established in the mid-nineteenth century in which women, mostly from the middle class, were able to escape from uncongenial family life into careers of devotion and good works. Each sisterhood was firmly placed under the control of a clergyman as Warden and a Bishop as visitor. Although Deaconesses were appointed later in the century they were not allowed to administer Communion or give the Blessing. In contrast to the Salvation Army, women could not become Captains in the Church Army. When women obtained the Parliamentary vote in 1918, however, there began to be a demand for them to have greater recognition within the national Church and in 1919 they were enabled to become members of its National Assembly. A few liberal churchmen, including William Temple, saw no reason of principle why they should be permanently excluded from Holy Orders but there was a strong feeling that nothing should be done which would damage relations with Rome. Before 1939 only a very few churchmen, such as Dean Inge of St Pauls and Canon Charles Raven of Cambridge, openly advocated the ordination of women, which was rejected by the Lambeth Conferences of 1920 and 1936.

After the Second World War, throughout the long debate on the ordination of women in the Church of England, which continued from the 1950s to the 1980s, there was always in the background the inescapable fact that in 1944 the Bishop of Hong Kong, R. O. Hall, had ordained a woman to be a priest. One side held fast to the precedent. The other dismissed it because the Bishop had been rebuked and the woman had been made to give up her ministry as a priest. An irony was that the Bishop had not in principle been a supporter of the ordination of women.

Ronald Owen Hall (1895–1975) was, says his biographer, sometimes a burning and shining light in a dark and uncertain world, sometimes a maddening autocrat, often controversial, but a saint.[14] He was the son of an Anglican clergyman; two of his brothers were missionaries and another a priest. In the war of 1914–18 he served in the infantry, won an MC and bar, and was the youngest Brigade Major in the Army. After the war he returned to Brasenose College, Oxford, was ordained and worked for the Student Christian Movement. His experience after this was quite different from that of most future colonial bishops. He served with the YMCA in Shanghai under the leadership of a Chinese whom he greatly admired. The Chinese people had opened his heart to friendship, he said, after the horrors of the Great War. He developed a highly unusual concept of the role of a missionary. He advised anyone aspiring to become one in China to sell every book on the religion of the country and to buy instead some fragment of Chinese art and to make friends with artists in order to discover the secret of saints. Training colleges, he said, should exchange their psychology books for violins and replace the professor of ethics by a teacher of art, for the Chinese were an artistic rather than a religious people and only by understanding the relevance of art to their life could the relevance of religion be understood.[15] He later served for a few years as vicar of a poor parish in Newcastle where his conviction was shaped that the most important object of his life was to relieve the sufferings of the poor.

There was some surprise when at the age of 37 he was appointed Bishop of Hong Kong. At his enthronement he asked for prayers that 'I may not say that a man or woman is no good but that rather on my knees I may say "I have not understood".'[16] A disciple of Maurice and of Gore, he habitually championed the underdog. Nettled by his criticisms of the administration from the pulpit of the Cathedral, the Governor referred to him as the Pink Bishop.

When the Japanese took Hong Kong in 1942, Hall happened to be in America. On his return he lived in other unoccupied parts of his large diocese, which included most of South China. In 1943 he wrote to the Archbishop of Canterbury, his friend William Temple, to inform him that he had given permission to deaconess Florence Li Tim-Oi in Macao to celebrate the Lord's Supper. The small island of Macao was a colony of Portugal, which was not at war with Japan. It could no longer be reached by priests from Hong Kong and only with difficulty from the mainland. The deaconess, Hall explained, was a graduate of a theological college where she had passed with honours the full course taken by the clergy. She had taken charge in Macao of a congregation of refugees from occupied China composed largely of school masters, university lecturers, and former government officials. Remarkably successful, she 'had developed as a man-pastor develops, and has none of that frustrated fussiness that is noticeable in women who having the pastoral charisma are denied its full exercise in the ministry of the Church'. 'I am not an advocate of the ordination of women,' he explained, 'I am however determined that no prejudices should prevent the congregation committed to my care having the sacrament of the Church.' Buried in this long letter, which also dealt with the need for British sympathy with Chiang Kai Shek, was a sentence whose significance Temple perhaps did not fully appreciate, 'If I could reach her physically I should ordain her priest rather than give her permission [to celebrate Communion], as that seems to me more contrary to the tradition and meaning of the ordained ministry than to ordain a woman.'[17]

The Archbishop replied saying that the matter appeared to be one for Hall's Province, but that in his judgement it was preferable to commission deaconesses to administer Holy Communion rather than to ordain them. Such permission could be withdrawn at the end of the emergency, whereas ordination would be permanent. Before receiving Temple's reply, Hall managed to meet Li Tim-Oi on the mainland and ordained her as a priest. In reporting this to the Archbishop he explained,

> Please be sure that my reason was not theoretical views of the equality of men and women but the needs of my people for the sacraments.... I have had an amazing feeling of quiet conviction about this – as if it was how God wanted it to happen rather than a formal regularisation first which could result in women who claim the right to be priest pressing into ordination even when there was no need for them as priests.

He proposed to report what he had done to his General Synod after the war and to press for provincial action to regularise the ordination of women in similar cases.[18]

Temple replied withholding comment until the action could be considered by the other bishops in China. When the news leaked out, the *Church Times* heavily accused Hall 'of neither considering the wider implications of his action nor consulting wiser heads than his own. Sober judgement is unable to conceive how the Bishop of Hong Kong could place even the supposed spiritual interest of a single congregation before not only the prescriptions of the Church, but the unity of Christendom.'[19] This of course is just what Hall had consciously done. The pastoral need for which he was responsible was more important to him than 'the prescriptions of the Church and the unity of Christendom'.

Temple died suddenly in November 1944. In his office was found the draft of a letter to Hall saying that his action was contrary to all the laws and precedents of the Church. 'I therefore feel obliged to tell you that I do profoundly deplore the action you took and have to regard it as ultra vires.... I find it hard to believe that the House of Bishops in China will in the long run be able to regard the action of one of its members as a matter outside its own jurisdiction.' Not only had Temple not signed this letter, but it is quite likely that in his illness it had been drafted by a member of his staff and that he had not seen it. His successor, Archbishop Fisher, read out the letter to Convocation and endorsed it. He then asked Hall to suspend Li Tim-Oi. Hall was courteously defiant in replying 'ecclesiastically I know my action was ultra vires. Spiritually I know it was not.' He had advanced Li Tim-Oi to the priesthood, he added, 'with a very strong sense of Our Lord's approval'.[20]

When the Chinese House of Bishops met in March 1946 they were confronted with a letter from Fisher accusing Hall of 'obstinately refusing the advice of his friends to suspend the woman and await the judgement of the Church'. The bishops were asked to repudiate his action. Several of them were expatriates. All of them counted on financial help from the west in the enormous tasks of reconstruction in their dioceses after the war. They passed a resolution by 7 votes to 1, with Hall and two others abstaining, 'regretting the uncanonical action of the Bishop of Hong Kong in ordaining deaconess Lei to the priesthood and asking him to accept her resignation from her priestly ministry'. The Bishop's own Diocesan Synod supported him. They sent a circular letter to the other Diocesan Synods stating that they found the attitude of the Church in the west impossible to understand: 'We believe God is using China's age old respect for women, and traditional confidence in women's gifts to open a new chapter in the history of the Church.' They proposed that an experiment by which deaconesses could be ordained as priests be allowed for 20 years. The General Synod prudently passed on to the Lambeth Conference of 1948 the question as to whether such an experiment would be in accordance with the Anglican tradition and order. The Lambeth Conference replied that it would not, and that it would gravely affect the internal and external relations of the Anglican Communion.[21]

Meanwhile, Florence Li Tim-Oi went back to minister in Macao quite unconscious of the controversy which was going on about her until summoned by the Bishop's secretary, a Chinese, who told her that the bishops at Lambeth would not accept her ordination, and that either Hall must resign or she must forfeit her title of priest. Immediately she wrote to the Bishop. 'I would like to keep quiet and help the Church. I am a mere worm, a tiny little worm.' She did not need, she said, the title of priest in order to do church work. But in giving up the title, and the functions which went with it, she recognised the indelible nature of Holy Orders, which she never renounced.[22]

She was transferred to Hoppo on the Vietnamese border where she now had to send the elements to the Archdeacon for blessing. When the Communists took over China, she was made to work in factories and on a chicken farm. All her books, including her Bible and prayer book, were taken from her and burnt. She and Christian friends would worship in the fields, reciting biblical passages and prayers which they remembered. When the persecution was relaxed she emigrated to Canada to join her sister and ministered to the Chinese population of Toronto. In 1984 the Movement for the Ordination of Women brought her to London for a service to commemorate the fortieth anniversary of her ordination, which was attended by 10 bishops. The Archbishop of Canterbury sent her a message – 'Sometimes you have suffered from misunderstandings about your ministry. You have never been eager to promote yourself, but only to build up the life of the Church and serve its mission in places of desperate need. Your life is an example to us all.'[23]

Bishop Hall remained in Hong Kong until 1966, working vigorously to improve the social services. In retrospect he considered that it had been a mistake to ask permission to make the experiment of ordaining a woman priest. 'We should have said,' he reflected, 'we propose to let the Rev. Florence Lee exercise her ministry and will report again at the next conference on the progress of the experiment. We ask you not to excommunicate us.'[24] For many years after the 1948 Lambeth Conference Bishop Hall never spoke of the ordination of women. In 1958, however, he ordained Jane Hwang as a deacon and in 1962 Joyce Bennett. He retired in 1966 after 34 years as Bishop of Hong Kong.

Hall's successor, Gilbert Baker, had gone out to China as a missionary and been ordained by him. Unlike Hall he was a firm believer in the principle of the ordination of women but was methodical and less impulsive, described by his Chinese Archdeacon as 'like the traditional Chinese who has all the appearance of indecision and keeps his mind in a state of indifference to all'. At the Lambeth Conference of 1968 he put the case for ordination of women in his diocese. The question was referred to the newly established Anglican Consultative Council, which by 24 votes to 22, with the Archbishop of Canterbury opposing, voted that all Anglican provinces were encouraged to remain in communion with any in which women were ordained. With this authority Baker ordained as priests the two Chinese and British women who had been ordained as Deacons by Hall. The precedent was soon followed by several Anglican communions elsewhere and eventually by the Church of England itself in 1993.[25]

Whilst men in several Christian denominations can be seen to have played a courageous part in asserting the equality of women, it is interesting that Millicent Fawcett recollected that in her long campaign for woman's suffrage none of the three best men she had known, her husband, J. S. Mill and Henry Sidgwick, had been Christians.[26]

9 Parliament and the Suffrage

The substance of this debate will be carefully reported in the newspapers . . . and go to every town and village in the United Kingdom and to every English speaking country under British rule; and therefore we shall secure that at least one day a year there will be a general discussion on a question so deeply affecting the interests and privileges of a large proportion of Her Majesty's subjects.

Jacob Bright, introducing a motion for women's suffrage in the House of Commons (30 April 1873)

Fifty years were to pass between the first formal proposal in the Commons by J. S. Mill that women should be given the Parliamentary franchise and its acceptance. Though some of those who voted for it in 1867, including John Bright and Henry Labouchere, had only done so out of respect for Mill and opposed women's suffrage when he was no longer in Parliament, its supporters could at first be optimistic. Only three years later a resolution to the same effect proposed by the Liberal Jacob Bright was carried by 126 votes to 93, though it was killed in Committee. Similar resolutions on bills proposed by private members were to be passed in 1884, 1886, 1887, 1897, 1904 and 1908. They remained only gestures; each time the Government failed to follow up by introducing legislation to implement them.

Again and again MPs of all persuasions earnestly declared that this was not a party question, but in fact it was so. Neither of the ruling parties would take action because in each of them the convictions of the leadership were different from those of the rank and file. Whilst in several Parliaments the majority of the Liberals were pledged to support women's suffrage in some form, three of their Prime Ministers in the period between 1868 and 1916, Gladstone, Rosebery and Asquith, were firmly opposed. Only Campbell Bannerman, who was Prime Minister from 1905 to 1908, was mildly sympathetic, advising a suffragist delegation which visited him 'to go on pestering'. He did not however give the cause sufficient priority to include it in his party's crowded legislative programme after its long years in the wilderness. Further, although a higher proportion of Liberal than Conservative MPs favoured women's suffrage in principle, some of them were lukewarm or even hostile to proposals which would only enfranchise women householders and ratepayers on the same basis as men, because this would exclude most working-class women and probably favour the Conservatives.

Among the Conservatives the position was the reverse. Their Prime Ministers, Disraeli, Lord Salisbury and A. J. Balfour, were in favour of women's suffrage, as was Bonar Law, their leader between 1911 and 1918. None of these however felt strongly enough to seek to convert their followers and to risk a revolt from the country backwoodsmen and many other of their followers. Yet there were a few Conservatives who joined with Liberals or even took the lead in proposing the almost annual motions in favour of women's suffrage. Thus when Jacob Bright, who had proposed the motion in 1870, 1871, 1872 and 1873, was temporarily out of Parliament, the Conservative William Forsyth took over the Bill in the three following years. Forsyth, however, angered many Liberals by insisting that married women should not be included on the grounds that to do so would virtually give two votes to their husbands. The Liberals Leonard Courtney and William Woodall took the lead in the eighties, but it again passed to the Conservative members Sir Albert Rollitt and Faithful Begg in the nineties. These annual debates became a ritual exercise. They gave a focus to the activities of the constitutional women's suffrage societies under Lydia Becker and later Millicent Fawcett; but in a crisis, when the party whips were put on or there was an important tactical consideration, even those who were marked on the filing cards of Mrs Becker and Mrs Fawcett as 'known friends' were apt to forsake the cause. Thus a hundred Liberal 'friends' voted against an amendment to Gladstone's Reform Bill of 1884 to enfranchise women when he made its rejection a matter of confidence. Perversely, the majority of Conservatives, though disliking the idea of women's enfranchisement, voted for this amendment in the hope of bringing Gladstone's Government down. Similarly the Irish Nationalists voted against women's suffrage proposals in 1912 and 1913 not on their merits but because they feared that if they were passed Asquith would resign as Prime Minister and might be replaced by someone less committed to Irish Home Rule.

This chapter is mainly concerned with those Parliamentarians who were not merely sympathetic to women's suffrage but were prepared to give it high priority even, if necessary, to the disadvantage of their careers. It may first be useful to summarise very briefly the process which led to women obtaining the Parliamentary vote. Ever since Mill's time various societies had agitated through constitutional means for this. They came to be consolidated in the National Union of Women's Suffrage Societies (NUWSS), led by Mrs Millicent Fawcett. In 1903 however, the more militant Women's Social and Political Union (WSPU) was formed by Mrs Emmeline Pankhurst and her daughters. The Pankhursts' indignation was fired to action when from the gallery of the Commons they witnessed a Private Member's Bill proposing women's suffrage being deliberately talked out. When the Liberal Government took no action, after their landslide victory of 1906, despite the commitment of most Liberals to the cause, the WSPU organised demonstrations outside Parliament and heckling of ministers. Later its campaign extended to destruction and damage to post boxes, shop windows, public buildings, art galleries, private houses and even churches, as well as physical assaults on ministers.

In 1910 an All Party Conciliation Committee, chaired by the Conservative Lord Lytton, obtained broad agreement on a Bill which would enfranchise only women who were householders or wives of householders. A Private Member's Conciliation Bill along these lines was passed by a majority of 110 in 1910 although the Cabinet Ministers Lloyd George and Winston Churchill opposed it as undemocratic. Reintroduced after the General Election of that year, with amendments to meet the criticisms, it was again passed handsomely by a majority of 167. The Government still took no action. By the time the Conciliation Bill was introduced for the third time, early in 1914, many Liberals and Conservatives who had previously voted for it opposed it because they felt that it would appear cowardly to seem to be yielding to the increasing violence of the Suffragettes, and it was defeated by 14 votes.

Asquith now introduced a broad Reform Bill to enfranchise the remaining men who had no Parliamentary vote. He agreed to allow amendments to be moved to include women in the Bill, indicating that they would be accepted by the Government if carried. The Speaker of the House, however, ruled that if such an amendment were carried it would so alter the Bill that under Parliamentary procedure it would have to be withdrawn. The whole Bill was then withdrawn and the Government's only action was to offer time for a Private Member's Bill, which experience over the past 40 years had shown to achieve nothing whatever. Increasing violence continued until the outbreak of the World War in 1914, when the Suffragettes suspended their protest and supported the war effort. A general reform of the franchise was now seen to be imperative; most men were qualified to vote by virtue of being householders who had to have been in occupation for the 12 months prior to an election; millions of men who had been absent while in the armed services would thus be disqualified in a postwar election. An All Party Conference chaired by the Speaker in 1916 recommended a broad measure of suffrage for almost all male adults. The suspension of militancy, and the contribution to the war of women who had maintained the economy by replacing the men who were at the front, now brought agreement that they too should be enfranchised on the same qualification; this was only to be at the age of 30, however, instead of 21 as in the case of males and 18 in the case of men in the services, in order that women electors should not outnumber men. These proposals were passed by such an overwhelming majority in the Commons that they were reluctantly accepted by the Lords. Women were thus enabled to vote at the age of 30 and to become Members of Parliament at 21 in 1918. Ten years later the age limit for women voters was made the same as that for men.

LIBERALS

There were always a few Liberals who were prepared to put the cause of women's suffrage before their careers. Their most striking gesture occurred in 1884 when three ministers, Sir Charles Dilke, Henry Fawcett and Leonard Courtney, refused to vote against William Woodall's amendment to include women in the Third Reform Bill, which greatly extended the franchise in the rural areas. Gladstone as Prime Minister refused to accept the amendment, ostensibly on the grounds that the Bill was already so bulky that to load it further risked rejection in the House of Lords; the instruction which he gave to his followers to oppose the amendment was made despite a hint from the Conservatives that it would not be unacceptable to the Lords.

The three rebels were all graduates of Cambridge. Henry Fawcett (1833–84) qualified as a barrister before becoming a Fellow of Trinity Hall, Cambridge, and Professor of Political Economy in the University. When he was elected to Parliament in 1865 he continued to hold his Chair. At the age of 25 Fawcett had been blinded in a shooting accident. The cheerfulness and ability with which he subsequently pursued both an academic and a political career made him famous and popular. His interpretation of the principles of laissez faire gave him a hatred of all restrictions on the rights of the weak in the interests of the strong. This had caused him to be a principal promoter of the Second Reform Bill of 1867, which extended the suffrage to the urban working class. His economic principles made him the only Member to oppose a Bill limiting the hours of labour of women, arguing that this would diminish their opportunities in competition with men, whose hours of work were not similarly limited. On the other hand, as Postmaster General he brought women clerical workers into the Post Office, which became by far their largest Government employer.

If Fawcett's Cambridge pupils brought him a problem, he used to ask 'Have you looked it up in Mill?' When in 1867 he spoke in favour of the latter's proposal to extend the suffrage to women he commenced by saying that all his lessons of political life had been learnt from Mill. His main contribution to the debate in that year was to tell the House that the experiment at Cambridge University had already shown that if women were given the same opportunities as men their characters were not debased but strengthened. In subsequent debates Fawcett ridiculed the argument that women would neglect their domestic duties if enfranchised, by asking whether male doctors, barristers and artisans were considered unable to study public questions without neglecting their professions. Women, he insisted, needed to be directly represented in Parliament so that their views could be known on questions which concerned them. Notable examples of these were their exclusion from trade unions and from the benefits of educational endowments managed by men who decided that the needs of all boys should be satisfied before those of girls could be considered. In general Fawcett followed Mill in maintaining that the question of whether women were intellectually equal to men ought to be determined by giving them the same opportunities as men and not by preconceived principles, religious or secular.

Fawcett, though nominally an Anglican, was an agnostic like his mentor Mill and his closest academic colleague and biographer, Leslie Stephen. He died suddenly at the age of 50, but his influence continued to be extended through his considerably younger wife, Millicent Garrett Fawcett, who had been his amanuensis and collaborated in his academic as well as in his public work. After leading the constitutional women suffragists for many years she was to live to see his ambition of Parliamentary votes for women achieved.[1]

Leonard (later Lord) Courtney (1832–1917) was an exact contemporary of Fawcett and like him combined the teaching of economics with a Parliamentary career. After graduating brilliantly at St John's College, Cambridge, he worked on the staff of *The Times* and as Professor of Economics at University College, London, where he allowed women to attend his classes even before they could become members of the College. He entered Parliament in 1876 where he came to inherit Mill's unofficial position as 'the conscience of the House'. His advocacy of women's suffrage was an expression of a broad concern for the oppressed and under-represented which caused him to resign from Gladstone's Government in 1884 when it failed to incorporate proportional representation in the Reform Bill, and later to lose his seat for opposing the Boer War.

He based his speeches on behalf of women's suffrage in Parliament on a Utilitarian philosophy. In introducing the annual Bill in 1878 he argued that 'by advancing woman and making her fuller, freer and nobler you will advance man with her'. Over many years Courtney asserted that to accord women's suffrage would be a demonstration that women were full citizens, and would improve both their own characters and that of Parliament through the interest they would take in the conduct of those elected. His other main argument, like that of Fawcett, was that women had a special interest in Parliamentary legislation on such current questions as temperance, labour conditions, education and divorce. Courtney broke with the Liberals over Irish Home Rule and became a Liberal Unionist in 1886. He was then disqualified from active politics by becoming Deputy Speaker from 1886 to 1892. Like Fawcett he was a philosophic Radical and disciple of Mill, pledged to laissez faire, free trade and religious liberty. He became a peer in 1906 and continued to support women's suffrage in the House of Lords until it was granted.[2]

Sir Charles Wentworth Dilke (1843–1911) was younger than Fawcett and Courtney. He came from a wealthy and aristocratic family and inherited a baronetcy at the age of 21. After spending much of his boyhood in France he entered Trinity Hall, Cambridge, where Fawcett was his tutor and intellectual mentor. He was elected MP for Chelsea at the age of 25. His French connections influenced him to declare himself a republican, and he was proud of his family's descent from Thomas Wentworth who had died in prison for opposing the personal rule of Charles I. He regarded the enfranchisement of women as a step towards a complete democracy. With Jacob Bright in 1869 he succeeded in obtaining (or, as some maintained, restoring) votes for women in municipal elections.

After serving as Parliamentary Under-Secretary for Foreign Affairs he entered the Cabinet at the age of 39 and was spoken of as a future Prime Minister. Women's suffrage might have come earlier but for his downfall through involvement in a spicy divorce case in 1885 which resulted in his being omitted from the Cabinet. He lost his seat in the following year and though he sat in Parliament again from 1892 until his death he never held office again. Women's suffrage however remained one of his most important preoccupations. Over his long political career his attitude to this never changed. When in 1869 opponents maintained that most women did not want the vote, he told Parliament that a class which has always been excluded from power and governed despotically seldom asks for power. At the very end of his life, in 1910, he wrote a pamphlet in which he criticised the repeated attempts which had been made to obtain a franchise for women on the same basis as that by which men then qualified and which would have excluded married women and most working women. The only satisfactory solution, he wrote, had to be a complete adult franchise for both men and women, as in New Zealand. Although Dilke had been a feminist from his undergraduate days when he sat at the feet of Fawcett and first met with Mill, he was fortified in his views by his second wife, the widow of Mark Pattison, the Rector of Lincoln College, Oxford. Emily Dilke was an effective promoter of the improvement of the social and industrial condition of working women through their enrolment in trade unions, as well as an eminent writer on art. Lady Dilke indeed provided a striking example of the capacity of women which was under debate.[3]

Another Cambridge Radical feminist was James Stuart (1843–1913), Professor of Engineering in the University and Fellow of Trinity College, who succeeded to Fawcett's seat in the Commons on his death. He was not only an active suffragist but a pioneer of women's education in his 'peripatetic university'. He was also a close collaborator with Josephine Butler in her campaigns.

Alongside the Cambridge Radicals in the cause were the Nonconformists, notably Jacob Bright (1821–99) who after Mill lost his Parliamentary seat introduced his proposal for women to receive the vote, on the same qualification as men, on five occasions. Together with Dilke, as has been seen, he took the lead in obtaining votes for women in municipal elections. In his commitment to women's suffrage he differed from his better known brother John Bright, the most renowned Nonconformist politician and most brilliant orator of his time, who served in two of Gladstone's Cabinets. John Bright, though a Quaker, had a patriarchal attitude to women, who he hated to see assert themselves. When John spoke in Parliament in 1876 against a motion to give women the vote, Jacob delivered a devastating attack on him at a meeting in the Manchester Town Hall. John had maintained that Parliament satisfactorily represented those who had no votes as well as those who did: Jacob replied that Parliament, which was always pressed for time, in fact only responded to those who could exercise the most pressure on it. He mocked John for bolstering his case with such feeble examples as that women were more favoured than men because they could sue for breach of promise, that fewer of them were hanged, and that there was a small tax on male but not on female domestic servants. To John's objection that women voters would be influenced by the Church, Jacob replied that the existing influence on male voters of landowners, mill owners and brewers was far worse. He persistently kept the issue before the public with examples of petty discrimination such as that of the Royal Agricultural Society, which refused to analyse fertilisers or soil samples submitted by women farmers though it would do so for men.

Jacob Bright, like John, after a brief Quaker education had entered the family spinning business. He became Mayor of Rochdale at the age of 35, then MP for Manchester. With a brief interval from 1874 to 1876, he continued to sit in Parliament until 1895, remaining loyal to Gladstone in 1886 when his brother John left the party over Irish Home Rule. A *Vanity Fair* cartoon described Jacob as 'The Apostle to the Women' and a Manchester newspaper commented that he was more Ursula's husband than John's brother. Ursula Bright, as treasurer of the Women's Property Committee, had a considerable responsibility for the passing of the Married Women's Property Acts of 1881 and 1882. Again, as secretary of the Women's Franchise League she played an important part in the inclusion of votes for married women in the Local Government Act of 1894. Jacob and Ursula were partners in many feminist enterprises, including the Contagious Diseases Campaigns. Ursula was the lobbyist; their close friend Richard Pankhurst drafted legislation; and Jacob introduced it in Parliament. Recognised as totally without political ambition, he never held office though he was made a Privy Councillor on retirement. After his death his daughter inspired his grandson Sir Charles MacClaren, a journalist, lawyer and industrialist, when he became a Liberal MP, to continue to introduce Jacob's annual resolution. One of Jacob's collaborators with a similar background was William Woodall (1832–1901). A partner in a china manufactury in Burslem and MP for Stoke-on-Trent and Hanley, he incurred Gladstone's great displeasure when he refused to withdraw his amendment to the Reform Bill of 1884 proposing that it be extended to include women, and he also introduced Women's Franchise Bills in 1887 and 1888.[4]

The Unitarian James Stansfeld was as consistent a supporter of women's suffrage as he was of their entry into the universities and into medicine. He insisted, like Jacob Bright, that it was only when a class received a vote that it gained a chance in the competition for precedence in legislation. He warned the House that by opposing a modest measure of enfranchisement on a household basis members might precipitate the result which they most apprehended – universal suffrage. Stansfeld however was conscious of the unpopularity he had acquired in his campaign against the Contagious Diseases Act and felt that it was better for the women's suffrage cause that its leadership should be in the hands of other Members of Parliament.[5]

As long as Gladstone was their leader the Liberal suffragists could make no progress. Conscious that they were becoming a majority in the Parliamentary party, he was too wily to declare outright hostility, frequently stating that the question required further consideration or was inopportune. Only in 1892, almost at the end of his political career, did he come out in the open in a letter, intended for publication, to Samuel Smith, MP, who had asked his advice on how to vote on a motion about to be proposed to extend the franchise to women on the same qualification as men. Gladstone declared that the suffrage would 'bring about a fundamental change in women's providential calling despite the difference in type between the sexes ordained by God'. He feared that it would lead to a dislocation of family life and that women 'would be invited unwittingly to trespass upon the delicate purity and refinement which were the sources of their power'. He hoped, he said, that the House of Commons would decline to give a second reading to the Women's Suffrage Bill. The majority of Liberals followed his advice and the motion was lost.[6]

There was opposition not only from the socially conservative Gladstone but among some Liberal intellectuals who were influenced by the Positivists, whose founder, Comte, taught that women should be adored but not treated as colleagues. Among these was the influential editor, writer and member of Gladstone's Cabinet, John Morley. Another Liberal intellectual opponent who became a Cabinet Minister was the Scottish Presbyterian James Bryce, Professor of Law at Oxford. Although he did as much as anyone for higher education of women, he opposed women's suffrage because women were 'not politicians, knew little and cared less about political questions'.[7]

When the Liberals came back into office in 1905–6, not only the bulk of their MPs but three-quarters of the Cabinet were committed to women's franchise. Sir Henry Campbell-Bannerman (1836–1908) was Prime Minister from 1905 to 1908. Speaking on a Private Member's motion to enfranchise women, in 1907, he said that this was a question on which there was a difference of opinion in all parties, on which the division should be left to the House; personally, however, he was in favour of giving the Parliamentary vote to women who paid taxes but had no direct representation, and who had to obey laws which they had no part in shaping. Whilst he voted for a Bill which would give the franchise to women on the same basis as men, he regretted that this aimed only at enfranchising that minority of women who were single or well to do. Campbell-Bannerman was a Scottish businessman, educated at Glasgow University and Trinity College, Cambridge, who took the advice of his shy, stout wife Charlotte on all important decisions. Although he did not give a high priority to women's suffrage and was irritated by the tactics of the Suffragettes, had he lived longer his influence might have tipped the balance in the Cabinet in favour of earlier legislation and thus the most violent phase of the campaign might have been avoided.[8]

His Chancellor of the Exchequer, H. H. Asquith, who succeeded him and was Prime Minister from 1908 to 1916, was firmly opposed. However, among the strongest supporters in principle in his Cabinet were Lloyd George, who successively held various senior posts, including that of Chancellor of the Exchequer; the Foreign Secretary, Sir Edward Grey; and Richard Haldane, Secretary of War and later Lord Chancellor. David Lloyd George (1863–1965) was an outsider to the Establishment. His father had died when he was in the cradle and he had been brought up by an uncle who was a shoemaker – most radical of all nineteenth-century professions. He had no higher education and had practised as a solicitor before entering Parliament. A successor to the earlier radical advocates of women's suffrage, he is said to have been converted to the cause by reading Ibsen's *Hedda Gabler* and *A Dolls House*. He spoke frequently and eloquently on the subject. 'You cannot trust the interests of any class entirely to another', he declared, 'and you cannot trust the interests of one sex to another sex.' There was no doubt about Lloyd George's commitment to the cause, but there was a question as to the tactical priority which he gave it. He became more and more uneasy about proposals such as those of the Conciliation Committee, which seemed likely to add upper and middle-class women but not working-class wives to the voters' register. Also, though he had made more speeches than any other minister in favour of women's suffrage, he became an indignant target of the Suffragettes, who broke up his meetings and physically attacked him because the Government had taken no action. When he succeeded Asquith as Prime Minister of the wartime Coalition Government in 1916 his way was clear. The Suffragettes had dropped their campaign and energetically supported the war effort and he led a coalition of Conservative, Labour and some Liberal Ministers most of whom were committed to women's suffrage. Even Asquith, now leader of the Opposition, announced his conversion in 1918.[9]

Sir Edward Grey (1862–1933), later Viscount Grey of Falloden, like Asquith was educated at Balliol College, Oxford, and served as Parliamentary Under-Secretary for Foreign Affairs in Gladstone's administration of 1892 and Foreign Secretary from 1905 to 1916 in the administrations of Campbell-Bannerman and Asquith. Grey came from a Whig family with a long tradition of public service and as reformers. His father died when he was 12 and he inherited the baronetcy from his grandfather whilst still an undergraduate. There seems to have been no notable maternal influence. He married when young and was devoted to his wife, who shared his preference for country life and disliked political society but seems either to have converted him to or strengthened him in his suffragism. She was killed in an accident shortly after he became Foreign Secretary. Because of the weight which he carried as a politician of unusual integrity his support was important. It is remarkable that in 1912–14, when foreign affairs were in the complicated and dangerous situation which led up to the First World War, Grey spared much of his time trying to obtain inter-party agreement at home on the almost equally intractable question of women's franchise. He countered the assertion that women were politically ignorant by maintaining that women who make homes are in no way politically inferior to the men who work to support them, and that the five million women who were earning their own living were as politically experienced as men. It was somewhat curious however that the minister who was in charge of a foreign policy which was about to result in the bloodiest war in British history should have dismissed the contention that women should not have votes because they knew little about foreign affairs, by saying that foreign affairs were not a significant issue in elections.[10]

Grey's closest political friend, Richard Haldane (1856–1928), came from a Scottish Nonconformist background, against which he revolted and studied at German universities. After achieving considerable success as a barrister he entered Parliament in 1885. In the Commons he sponsored Women's Suffrage Bills in 1889, 1890, 1891 and 1892 and did much to ensure that the provincial universities which sprang up at the end of the century admitted women. His biographer ascribes his active feminism to his deep respect for his mother who, after the early death of his father, lived to be nearly 100. Though an able administrator he was not a popular politician. His German education, for which he often and rather unwisely gave thanks, caused him to speak with a certain detachment from British prejudices. 'I believe', he ruminated in a Parliamentary debate in 1910 on the enfranchisement of women, 'that the time will come when people will feel that their doubts were the outcome of a great superstition and marvel that humanity had not emancipated itself earlier.' As Lord Chancellor his support was important but he was dropped from the Government, after a scurrilous campaign against him as pro-German, before the success of the cause.[11]

LABOUR

The nineteenth-century Radical tradition descended in the early twentieth century not only to the rejuvenated Liberals but to the Labour Party, which from 1906 began to be substantially represented in Parliament. Unlike the other two parties it had an official policy of support for women's suffrage. It was sharply divided however between those who would accept this initially on the same basis as the current male suffrage and those who insisted that it should only come as part of a new franchise which would be extended to all adults. Whilst for most Labour Members of Parliament this was only one among many important issues, three made it their principal cause.

The most ardent of these was George Lansbury (1859–1940) who sacrificed his seat in Parliament for it. He came from the working class, earning his living in a sawmill and as a carter. In a period of harsh unemployment he emigrated briefly to Australia where he found conditions even harder. He returned to live in Bow in East London and was an activist successively in the Liberal Party, the Independent Labour Party and the Labour Party. His first shift of allegiance was partly due to the reluctance of the Liberals to give priority to votes for women. 'There were certain things I always felt,' he said. 'How working women's life was an unending labour, cooped up in a little brick house. Boys could have evenings to themselves. Girls must work in the house and wait on them.' This seemed to him grossly unfair. He came to regard the women's vote as an essential weapon in the fight against capitalism. At first he argued that women should be given Parliamentary votes and enabled to become members of local authorities mainly because there were certain questions to be settled in which they were more expert than men. Later however, he came to condemn the notion of separate spheres and to link the women's and socialist movements, believing that only in a socialist society could women, half the human race, realise their full potential, and be treated as comrades. Lansbury became a warm supporter of Mrs Pankhurst and urged the Suffragettes to 'burn and destroy property and anything you like'. In 1912 he left the Parliamentary Party because it would not support the Suffragettes fully, and resigned his seat in the Commons: he failed to recover it in the subsequent by-election, in which he ran as an Independent. He was jailed for inciting women to criminal acts but was released when he went on hunger strike. When he returned to Parliament in 1922 the suffrage had been won. Lansbury was an enthusiastic supporter of the birth control movement at a time when his party distanced itself from the question.[12]

James Keir Hardie (1856–1916) had started work as a miner in Scotland at the age of 10. Like Lansbury he was a Christian Socialist, supporting the cause because he saw women as the bottom dogs, victims of capitalism, miserably paid and exploited at home. He was prepared to accept household suffrage as a first step, although most of his party considered that this would be against the interest of their class. When he was outvoted on this point by the Party Conference, he threatened to resign from the Parliamentary Party, of which he was leader. It was then agreed that Labour MPs would be allowed a free vote on the question. Friendly with the Pankhursts, and indeed probably Sylvia's lover, he now went so far as to propose that Labour should vote against every Government measure until women's suffrage was won. He failed to carry the Party on this. Hardie saw the vote as a means to freedom for women, the root of all of whose grievances was dependence on men. It was an economic and social freedom to be granted in exchange for duty to the community; a freedom to escape from being used to beat down the wages of men. Economic independence, he also suggested, would lead to freer choice of mates; the tendency would be for the less fit to get left out and the most fit to be taken, so that the race would be improved. He did not live quite long enough to see victory for the cause.[13]

Philip Snowden (1864–1937), though a weaver's son and self-educated, unlike Lansbury and Keir Hardie moved into the middle class, working as a clerk in the Inland Revenue until incapacitated by spinal tuberculosis. His devotion to women's suffrage was inspired by his marriage to a formidable school teacher, Ethel Anniken. Like Lansbury, he moved from the Liberal to the Labour Party, shocked by the conditions which caused 70 per cent of the weavers' wives in his home district in Lancashire to have to work in mills and factories. Entering Parliament in 1906, his contribution in debates was less emotional than that of Lansbury and Hardie, refuting with statistics arguments that women had no grievances which Parliament had not met, and pointing out their unfair treatment in national insurance and old age pension schemes. He was

Joint Secretary of the Conciliation Committee. His powerful oratory at the Labour Party Conference of 1913 persuaded it to resolve, despite the opposition of the miners, to oppose any Franchise Bill in which women were not included. Snowden was a Methodist who told the House of Commons that the franchise was a great moral question. He based his arguments not so much on women's fundamental rights as on the harm to the community if any class or section had to obey laws but had no responsibility for legislation. Snowden was still in Parliament when the Parliamentary franchise was extended to all women in 1928, and later he became Chancellor of the Exchequer.

In addition to the contribution of Keir Hardie, Lansbury and Snowden, that of Arthur Henderson (1863–1935) proved to be of considerable importance when, as the Labour Party's representative in the Coalition Government in 1917, he made it clear that he would support no extension of the franchise which did not include women. Ramsay MacDonald also advocated women's suffrage, not because it would benefit women but because it would benefit the state.[14]

CONSERVATIVES

Right through the 50-year period during which women's suffrage was debated in Parliament the leadership and a substantial aristocratic element of the Conservative Party were in its favour. Benjamin Disraeli, later Lord Beaconsfield (1804–81), who had led the party from 1867 to 1880, had been widely regarded as a clever adventurer without fixed principles. After his death however, the sanctified memory of Lord Beaconsfield was often invoked by Conservative suffragists. Disraeli, as an outsider to the Establishment who had escaped formal education at Public School and Oxbridge, had few conventional prejudices, and in the conversation and correspondence with women which he much enjoyed treated them as political equals. He had long held that in a country governed by a woman, and in which women could own land, hold legal courts and serve as church wardens, there was no reason why they should not have the right to vote. He allowed a free vote to his party on Mill's proposal to enfranchise women in 1867. But, having witnessed the fate of Peel after introducing policies unpalatable to his followers, he took no overt step to advance the cause of women's franchise.[15]

Although his successor Robert Cecil, Lord Salisbury (1830–1903), who led the Party from 1881 to 1902, told the Primrose League in 1886 that personally he favoured the women's franchise, he too did little to promote it; almost all the members of his family and their connections the Balfours, however, were ardent suffragists. His wife came from a Unitarian family. The Salisburys brought up their sons and daughters to exercise their wits, stimulating them to find fallacies in everything except the Church of England and to delight in exposing flaws in political slogans. Two of their sons, Lord Robert and Lord Hugh Cecil, championed women's suffrage in Parliament; another keen supporter was Lord Selborne, the husband of their daughter Maud, who herself was President of the Conservative Women's Suffrage Association.[16]

For one son, Lord Robert Cecil (later Viscount Cecil of Chelwood), women's suffrage was a principal cause before his energies were diverted to that of the League of Nations. He told the House of Commons in 1913 that women had been so useful in local government that the onus was on the other side to show why they should be excluded from being heard in national debates on such current issues as Irish Home Rule and the Education and Housing Bills. He challenged any Member to point to instances when women had used political or quasi-political power worse than men. As a Conservative he believed that whilst violent change was dangerous, resistance to change was even more dangerous. Lord Robert made two particularly important practical contributions. Firstly, like Arthur Henderson he only agreed to accept office in the wartime Coalition Government on condition that women were included in any Franchise Bill. Later as a minister he was able to insist that the Covenant of the League of Nations should provide that all posts in the League Secretariat should be open to women.[17]

Although almost all the Cecils and their Balfour relatives were suffragists, the member of the family who had most potential influence on the question after Lord Salisbury was his nephew and successor as Prime Minister, Arthur Balfour (1848–1930). Arthur was educated at Eton and Trinity College, Cambridge. His father died when he was a child and he inherited a fortune. He spent much of his childhood and youth amidst the Cecil family and that of the Lytteltons. Balfour entered Parliament at the age of 26 and early in his career was rebuked by an uncle, Beresford Hope, the Member for Cambridge, for proposing that women should be allowed to take university degrees. He rose rapidly to become leader of the Conservatives in the Commons. In 1892 he supported a motion for women's franchise, telling the House that a great deal of cant was talked about the danger that involvement in politics would degrade women, although at present MPs of all parties were delighted to use their services to help them to get elected. He also ridiculed the argument of some of the Imperialists in his party that women should not be enfranchised because they did not bear arms, pointing out that a high proportion of the corpulent and elderly gentlemen sitting beside him were equally unable to do so. He prophetically stated that in any further alteration of the franchise, 'the question would again arise, menacing and ripe for a solution'. His mild interest in the cause was kept under pressure by his two sisters-in-law Lady Betty and Lady Frances Balfour, who were active suffragists. He liked to visit and help Newnham College, Cambridge, whose Principal, his sister Nora Sidgwick, was pioneering higher education for women. In brief, he described himself as a steady but not vehement supporter of the women's suffrage movement, aware that his opinions were very unpalatable to many of his colleagues but pitched in too low a key to please ardent supporters. When after a long period as heir apparent he became Prime Minister, he did little to help the cause. Perhaps his most useful contribution to it was made behind the scenes later as a member of Lloyd George's Coalition Government when it introduced the all-party Bill in 1917 which conferred the vote on women.[18]

Two of Balfour's closest aristocratic friends and Parliamentary colleagues, Alfred Lyttelton and George Wyndham, were more outspoken supporters of women's franchise. Lyttelton (1857–1913), the son of a peer, was like Balfour a product of Eton and Trinity College, Cambridge. With the authority of a former Colonial Secretary he could remind those Conservatives who maintained that people in the Colonies would lose respect for a nation which allowed women to vote, of the enormous prestige, throughout the Empire, of Queen Victoria. Lyttelton was perhaps even better known as one of the most brilliant amateur cricketers in England than as a Cabinet minister. On the question of women's suffrage he spoke in the Commons with a sportsman's passion for 'fair play' shortly before he expired in a cricket match. 'It is weakness that deserves representation. The weak, the miserable, will be patient and more ready to forgive, even if you make mistakes, provided that they have had a chance of convincing you before you make those mistakes.' These, he insisted, had been the doctrines of Disraeli and Salisbury.[19] George Wyndham (1863–1913) was Balfour's Parliamentary Secretary before becoming Chief Secretary for Ireland. He was the brother of Balfour's closest woman friend, Mary, Countess of Wemyss, and was regarded by him almost as a son. He was educated at Eton and Sandhurst; one of his grandfathers was an earl and the other a baronet. Wyndham was a literary critic and poet who carried the patriarchal values of a country gentleman into politics as an Imperialist. In an eloquent appeal to the House in 1892 he said that if women acquired the vote, 'we should find even in the legislation which we so kindly introduced and passed on their behalf that we had wounded their feelings, missed their true welfare and lost in an evil day the guidance which we might have received at their hands'.[20]

An aristocratic supporter from an earlier generation in the Commons was Lord John Manners until he was translated to the Lords as Duke of Rutland. He had been a friend and close colleague of Disraeli from their early romantic phase, seeking to reintroduce into public life the chivalry of the Middle Ages. He would intervene to deplore the fact that women had no votes whilst their gardeners had them, and that women farmers were less privileged than their labourers. At a crucial point in 1884 when the amendment to include women in Gladstone's Suffrage Bill was debated, he hinted that the House of Lords would not oppose women's suffrage on this limited basis. Had Gladstone responded it is possible that British women might have gained the Parliamentary vote then, before those in any other country.[21]

Many of the Conservatives who favoured women's suffrage had been to Eton, in a regime which at the time was perhaps somewhat milder and more civilised than that of other Public Schools. Though the Cecils had gone on to Oxford, most of the other supporters were graduates of Trinity College, Cambridge. Their sisters were educated at home by excellent private governesses and were inspired by cultural travel on the Continent and sometimes by experience of social work. They were undeterred by too much grounding in the Classics, which instructed their brothers in the contempt for women in ancient Greece and Rome. It was from this circle that the Conservative Parliamentary leaders usually chose their wives. Among Conservatives who sponsored motions in the Commons in favour of votes for women were also some, however, with a middle-class professional or business background. Among them were William Forsyth, a barrister and graduate of Trinity, Cambridge, who introduced several Bills in the 1870s. Another was Sir Charles Rollitt, a graduate of London University, barrister, shipbuilder and Mayor of Hull. Faithful Begg, who introduced Bills in the 1890s, was a privately educated stockbroker and Chairman of the London Chamber of Commerce; his support was partly based on his experience of New Zealand where he had lived and married a local girl.

Perhaps the only Conservative whose devotion to the cause seriously jeopardised his political career was Victor, second Earl of Lytton (1876–1947). There was a strong strain of feminism in his family. His great grandmother Anna Wheeler had inspired William Thompson's early plea for women's suffrage, 'An Appeal of One Half the Human

Race'. His grandfather the novelist Edward Bulwer Lytton, whose life he wrote, had insisted that the mind of woman was equal to or superior to, that of man. Educated at Eton and Trinity College, Cambridge, he inherited the title at the age of 15. He found his cause when his sister Lady Constance Lytton became a militant member of the WSPU and lost her health as a result of repeated imprisonment. From 1909 for the next five years women's suffrage became his absorbing passion, horrifying the King, alienating his friend Churchill by the ferocity of his attacks on him, and damaging the prospects of a political career which might have carried him into the Cabinet. His work as Chairman of the Conciliation Committee however, nearly achieved success.[22]

THE IRISH PARTY

The Irish Party in the period consisted of some 80 MPs who at times held the balance of power between Liberals and Conservatives. In principle it maintained that the British legislation on the subject was of no interest because only an Irish Parliament would eventually decide whether Irish women should have votes. In practice it sometimes used its position tactically depending on its current relationship with the Liberals and Conservatives. There was one Irish Member however, the chivalrous Willie Redmond, brother of his party's leader, John Redmond, who took every opportunity to speak with emotion in favour of women's franchise, always making the same two points – first, that women were slaves and he was against slavery everywhere; secondly, that the granting of the vote to women in Australia had greatly raised the tone of political life. Both of the Redmonds had lived in Australia and married there.[23]

The sudden end to the controversy was partly due to the presence in the Coalition Government of 1916 of representatives of all parties who were strong supporters of women's suffrage. These included, as well as the Liberal Prime Minister Lloyd George, the Conservatives Balfour, Bonar Law and Lord Robert Cecil and the Labour Henderson. Eventually Curzon, the Foreign Secretary, though he was President of the National League for Opposing Women's Suffrage, advised his followers in the House of Lords, of which he was leader, to abstain and allow the Bill which enfranchised women to be passed.

There were, of course, several men outside Parliament at the time who made considerable sacrifices in the cause of women's suffrage. The most notable of these was F. W. Pethick Lawrence (1871–1961) who came from a Unitarian family. Graduating brilliantly at Cambridge he took a law degree and inherited a fortune before throwing in his lot with the WSPU as a gesture to prevent a sex war. With his wife Emmeline he edited its journal and spent much of his money in its support. As its treasurer he was held personally responsible for the damages done by the militants, fined, jailed and eventually made bankrupt. Mrs Pankhurst expelled the Pethick Lawrences from the Union when they wished it to be more democratic; Pethick Lawrence, however, when he later became a Labour MP for many years worked for women's rights in law and employment and for family allowances. Richard Pankhurst (1834–98), Mrs Pankhurst's husband, should also not be forgotten. A barrister, he gave much of his time to drafting Women's Suffrage Bills which came before Parliament, and to influencing the Independent Labour Party to commit itself to the equal status of women; he unsuccessfully ran for Parliament as an Independent with women's suffrage as a central point in his platform. It was he who inspired his much younger wife and her daughters to devote themselves to the suffrage movement.[24]

In 1913 *A Suffrage Annual and Woman's Who's Who* was published in which, though most of the entries were on women, men were also honoured. Most of the men then still living who are mentioned in this chapter are included. There are also several Church of England clergymen, including Bishop Hicks of Lincoln, as well as Nonconformist ministers and a notable representation of Jews, both MPs and rabbis. Writers include George Bernard Shaw, Israel Zangwill and H. W. Nevinson. There is a roll of male martyrs, including those jailed for, or injured in, heckling Lloyd George and Winston Churchill and for whipping the latter and breaking his windows. It is paradoxical that Lloyd George, the subject of some of these attacks, is also included as a friend of women's suffrage.[25]

It has been seen that the champions of the women's suffrage in Parliament were influenced by a variety of motives and experiences. Among them were Radical Nonconformists and Radical intellectuals, Socialists, aristocrats and middle-class professionals. They defy precise classification. The Bright brothers and again the Redmond brothers were on different sides. Eton produced not only suffragists but opponents as vehement as Labouchere and Curzon. Though the Unitarian Cabinet minister Stansfeld was one of the keenest feminists and advocates of the suffrage, his even more prominent co-religionist Joseph Chamberlain believed, and frequently said, that women had no role in politics, and forbade his daughters to discuss womens' rights. In all classes there were romantics and men whose devotion to the cause was inspired not by political principles but by relationships with mothers, wives, sisters and lovers.

Their efforts might have been successful earlier had they been unanimous about the qualifications by which women should be admitted to the suffrage. Though tactical political considerations were important in this, there were also genuine differences of principle. Some followed Mill in urging that women should be given the vote on exactly the same basis as men. This however would have excluded most married women because they were not householders, and thus appeared undemocratic and even unacceptable to many feminists. Some considered that women's suffrage should only come as part of universal suffrage, to which they were committed. Others felt that because there were a million more women than men in the country a difference in the qualifications of the sexes was necessary to prevent the interests of men being submerged. Few politicians appear to have changed their convictions on the question in response to arguments, though some altered their positions for reasons of party advantage or presumed advantage. Only towards the end of the long debate were there major changes – in both directions. In 1914 previous supporters felt obliged to vote against women's suffrage rather than appear to yield to the violence of the Suffragettes; and in 1918 previous opponents changed their view of the capacities of women in light of their proved ability to replace the men at the front in a wide variety of tasks. The agreement was reached in 1917–18 because the Liberals found themselves outflanked by Labour and saw a franchise for women over 30 who were householders or wives of householders as broad enough not to favour the Conservatives; whilst the latter feared that if left until after the war even more drastic reforms might be introduced and lessen their influence.[26]

10 Breaking into the Professions

A paralysing weight would be lifted from the mind of many a public servant could he feel a career of honourable usefulness was open to his daughter without involving loss of social position.

Anonymous article on 'Employment of Women in the Public Services in Europe' (*Quarterly Review*, January 1881)

Mill had been correct in predicting that once the main bastion of the Parliamentary franchise was taken the remaining ramparts of the Citadel which barred women from the professions and government services and maintained other inequalities would rapidly crumble. In the General Election of 1918, which it easily won, the coalition of Conservatives and Lloyd George Liberals pledged itself to remove all discrimination against women. In April 1919 the Labour Party introduced a Bill to remove disqualifications of women in civil and judicial appointments, which was passed in the Commons but defeated in the Lords. This roused the Government to introduce its own Sex Disqualifications (Removal) Bill. It met little opposition except in the Lords, which removed a clause that would have permitted peeresses to sit in their House. Under the Act passed in December 1919 it now became illegal for women to be excluded from the professions. Equality in government services was also promised but the Government was allowed considerable discretion in issuing regulations to bring this about. Thanks to the intervention of Lord Robert Cecil and J. W. Hills, Oxford and Cambridge, despite their statutes, were enabled, though not obliged, to matriculate women and admit them to their privileges.

Historically many of the professions had grown from mediaeval guilds, of which women were very seldom members, or from vocations in the Church, to which they could never aspire. In the nineteenth century associations had been formed within professions to improve the status of their members. These, though mostly starting as discussion groups, came to establish qualifications and organise training and examinations. They also concerned themselves with the reputation and emoluments of their profession. Several obtained Royal Charters and were authorised by Parliament to decide, and compile registers of, those who might be permitted to practise. At the time when they were established the idea that women could become members occurred to hardly anyone, but after 1878, when women were admitted into the medical profession and enabled to study at the universities, the issue began to arise. The councils or governing bodies of the professional societies tended to be controlled by older and more conservative members, uncomfortable, even shocked, by the prospect of a female presence, and arguing that only an Act of Parliament could bring this about, as had been necessary in the medical profession. The records of meetings and the articles and correspondence in the journals of the societies show how members who proposed the admission of women were usually successfully resisted, often right up to 1919 when Parliament settled the question finally. There was a great deal of self-interest at stake and the real reasons for opposing admission of women were often different from those cunningly, skilfully or fantastically deployed in public.

PHARMACISTS

One of the earliest debates took place among the pharmacists. In the early nineteenth century the medical general practitioners became increasingly jealous of the pharmacists who were taking away much of the dispensing of medicines, which had provided an important part of their income; they attacked them as being unqualified and dangerous. The pharmacists responded by forming a Pharmaceutical Society, which received a Royal Charter in 1843. Although there were a number of women already practising as pharmacists, none were admitted as members of the new Society. Indeed its founder, Jacob Bell, despite being a Quaker, treated the existence of a high proportion of women pharmacists in any part of the country as a symptom of the profession's backwardness there. In 1858 there was a catastrophe in Bradford where a druggist's assistant when mixing peppermint lozenges mistook arsenic for sugar and caused 20 deaths from poisoning. The incident led to legislation in 1868 prohibiting any persons to practise without being examined by the Royal Pharmaceutical Society. Parliament's use of the word 'persons' – whether it was deliberate or not – obliged the Society to examine and register women as well as men; but it refused to admit them to its lectures, its laboratories or its membership. Among the men who now registered were a number from the Provinces, more radical than the Londoners who had hitherto controlled the Society. One of these was Robert Hampson of Manchester who had been a medical student before an injury caused him to give up his studies and become a pharmacist. His wife was a midwife. The Hampsons believed strongly in co-education and together started a home for young women. Nothing, said an obituary of Hampson, moved him to anger more than the oppression to which women were subject. Such oppression he saw in the Society's treatment of them.[1]

In 1872 he moved that ladies be admitted to the lectures and laboratories of the Society, arguing that as Parliament had ruled that women were eligible to be examined and registered it was manifestly wrong to exclude them from the instruction necessary before being examined. He was supported by members who wanted their wives or daughters to become qualified to replace them when they went off shooting or botanising, and to carry on their business after their death. The Society agreed to admit women to lectures but not to its laboratories, ostensibly because there was insufficient space.

Women did outstandingly well in the professional examinations and Hampson in 1873 proposed that they be allowed to become members of the Society. The opposition to this was led by George Sandford who declared that 'there are more fitting occupations for ladies than listening to descriptions of bodily ailments over our shop counters. . . . I cannot help thinking the tendency of the present day is too much towards upsetting that natural and scriptural arrangement of the sexes which has worked tolerably well for 4000 years.' One of his colleagues supported economic with moral and physiological objections. Female colliers, fishmongers, clerks and doctors, he said, had removed landmarks and caused a universal depreciation of labour in both quality and remuneration. Women's constitutions were so delicate that when harrassed by disappointed affection or family affliction they might hand out false prescriptions; moreover, dealing with the most revolting afflictions could blunt their moral feelings.

When Hampson's proposal to admit women was defeated in the Council by the President's casting vote he carried the fight to the Annual General Meeting, which he told that the Society was no longer a private one and that the question was one of justice and fair play. Opponents put forward their customary argument that the work of the profession was too indelicate for ladies. The intervention of the Society's solicitor appears to have been decisive. After misleadingly comparing the Society with the Gentlemen's Clubs in London's West End, from which ladies were excluded, he advised that if they were admitted to the Society, members who objected could apply to the Courts to call upon it to justify its action and thus bring about very considerable legal costs. The motion for admission was lost by a large majority. But now the *Pall Mall Gazette* took up the cause in a series of articles accusing the Society of being in breach of the Pharmacy Act of 1868 in excluding women. It attacked it as dominated by tradesmen who were motivated by jealousy of competition in an industry in which there was a profit of 400 per cent to 600 per cent on prescriptions. It ridiculed it for refusing to allow women to compete for its scholarships and prizes on the grounds that they would have the unfair advantage of having more spare time for study than male students. Fortified by this outside support, Hampson again and again raised the question in the Council. It was only, however, when Parliament caused women doctors to be placed on the register of the Medical Council in 1878 that the arguments for prohibiting them from similarly being registered as pharmacists collapsed. When they were admitted to the Society, cold shouldered by male students and colleagues, they formed their own association of women pharmacists, whose members marched under its banner in Suffragette processions. By 1908 just 1 per cent of registered pharmacists were women. It was only in the 1914 war that a shortage of manpower led to a substantial increase in their numbers.[2]

ARCHITECTS

By the end of the nineteenth century the New Woman, armed with a university degree, liberated by the bicycle and inspired by Shaw and Wells, was knocking at the doors of many professions, most of which had become accustomed to a lowly female presence as clerks who had come in to operate the newly invented typewriters. In one profession they narrowly but rapidly achieved success. Although membership of the Royal Institute of British Architects (RIBA) carried considerable prestige, the Institute had no authority to prevent anyone practising as an architect without its qualification. A few women had indeed practised commercially from the late eighteenth century, mostly widows such as Elizabeth Deane (1760–1828) who completed the Naval Dockyards at Cork on her husband's death. There were also amateurs, aristocrats or wealthy evangelical ladies who built houses for their labourers or designed chapels, schools and almshouses. By the second half of the nineteenth century women too could be found in junior positions in architects' offices where their services proved convenient as neat and inexpensive copiers.

Only when the Married Women's Property Acts were passed were women generally enabled to make contracts to sue and be sued and thus to practise independently. Few did so except in the designing and decorative arts. Among these were Millicent Garrett Fawcett's sister Agnes and her cousin Rhoda who were befriended and taken on as pupils by the architect J. M. Brydon. He was eventually well rewarded for his feminist sympathies when Elizabeth Garrett Anderson arranged for him to be the architect of both the London Hospital for Women and the London School of Medicine for Women. Another friend of the women architects was Ernest George, who trained Ethel Charles in his office and, together with Brydon, nominated her as an Associate Member of the RIBA in 1898. Some other friends may have done as much harm as good whilst taking women into their offices but by their misguided chivalry confining them to separate rooms and shielding them from outside work, despite the daring innovation of divided skirts. In the 1891 census nineteen women were recorded as architects in England and five in Scotland.

In June 1898 Ethel Charles sat the RIBA examination successfully. The Institute's charter provided that its Associates should be 'persons' engaged in the study or practice of architecture. Relying on this, its Council accepted her nomination as an Associate. At the next General Meeting several members protested, pointing out as a precedent that whilst the Universities of Oxford and Cambridge and the Royal College of Physicians permitted ladies to sit for their examinations, they did not award them their degrees or diplomas. Professor Aitchison, the President, however, congratulated Ethel Charles on her initiative and spirit and observed that in light of public feeling RIBA would be very much behind the spirit of the age and would appear ridiculous if it refused to admit women. The nomination of Miss Charles was approved by 51 votes to 16. The opponents pursued the matter at the next General Meeting, proposing that no woman should be admitted until the opinion of a Queen's Counsel be obtained as to whether this was allowed under the RIBA's charter. The elderly proposer, a Mr Dawson, maintained that admission of women would provide an unfair competition to junior male members and that whilst ladies could be efficient assistants in the office they were quite unfitted for outside work. He was mocked by a member who sardonically enquired whether the septuagenarian Dawson could climb a ladder better than a young woman, and told him that his reasons savoured of a Sheffield trade union. A bizarre analogy used by another member to support admission of women was that they were deemed to be included whenever Church of England congregations were addressed as 'dearly beloved brethren'. The back-woodsmen had turned out in force and Miss Charles's nomination was only approved by the President's casting vote.[3]

VETERINARY SURGEONS

About the same time the Royal College of Veterinary Surgeons (RCVS) faced a request to admit women which, though refused, haunted it for 26 years. Aleen Cust was the daughter of a baronet with a post at Court and a mother who was a lady-in-waiting to Queen Victoria. After she announced her intention to become a vet her disgusted parents had nothing more to do with her; on entering the New Veterinary College in Edinburgh in 1896 she changed her name to Custance to avoid embarrassing them. Surviving the jeers of some fellow students who hurled anatomical specimens around the class, she applied in 1897 to sit for the first professional examination of the RCVS. A considerable discussion took place within the Society, in which letters to the *Veterinary Record* urging the admission of women were signed and those which opposed it were anonymous: opponents argued that women did not have sufficient physical strength and that it would be indecent and indelicate for them to carry out castration, foaling and calving. The RCVS took the advice of Counsel who observed that whilst the charter and Act under which it operated gave it authority to admit students, all students had been male at the time they were drawn up. He recommended therefore that the application be refused and that Miss Custance be invited to issue a writ against the RCVS in the Court of Queens Bench in London. The RCVS accordingly refused to examine Miss Custance because it was 'contrary to long usage and all precedent for women to enter the profession'. Aleen Custance could not afford to go to law but the Principal of the New College, William Williams, with his son who was a professor at it, sued the Society for damages in the Court of Sessions in Edinburgh. The RCVS successfully pleaded that a Scottish Court had no jurisdiction over an institution based on London, although three of the British veterinary colleges at that time were located in Scotland and appeared on the RCVS headed notepaper.

Aleen Custance remained at the New College for the necessary three years to complete the full diploma course of training, gaining a gold medal. She was also taken as a pupil to obtain practical experience by a veteran Dundee veterinarian who said that she was the best student he ever had. She then returned to Ireland, where she had grown up, to become assistant to William Byrne, one of Ireland's leading young veterinarians. Byrne was a feminist, inspired by George Meredith's novels, and he openly attacked the majority of the RCVS Council, of which he was a member, as 'misogynist old bachelors and henpecked husbands'. Aleen rode to hounds and her passion for and understanding of horses made her popular with everyone except the parish priest, who was shocked by a woman who carried out castrations but who as a member of the Protestant Ascendancy could ignore him. She was soon able to set up her own very successful practice among farmers and country gentry who cared nothing about diplomas issued from London. In 1905 the Galway County Council appointed her as its Veterinary Inspector. The RCVS Council was greatly perturbed at this appointment to an official post of a woman without its diploma and persuaded the UK Department of Agriculture to forbid it. The Galway County Council refused to appoint anyone else and the Department had to give way, only insisting that the title of 'Veterinary Inspector' be changed to 'Inspector'.

By now women were being trained as veterinary surgeons in France, Russia, America and Australia. One member of the RCVS Council, Professor Share-Jones, persistently drew its attention to the absurdity of a situation in which women could take degrees in veterinary science at Liverpool and London Universities but were not allowed to obtain the RCVS diploma which would enable them to practise, for Aleen Custance was unique in doing so without the diploma. The Council however, rigidly maintained that 'students' under its charter meant men. It rejected a request from another woman, Miss E. C. Knight, to be examined in 1919 after a passionate plea from its former president Sir John MacFadyean that it would be cowardly to yield to public opinion: at its next meeting however, it was informed by its secretary that under the Sex Disqualifications (Removal) Act women were now entitled to sit for its examinations.

As for Aleen Custance, she had briefly been engaged to an army officer serving in India but her fiancé concluded that she would not be tolerated by his regiment, 'who believed in feminine women'. Neither eventually felt able to sacrifice a career and the engagement ended. Aleen served in France in the 1914 war, first with the YMCA and then with Queen Mary's Army Auxiliary Corps, dealing with diseases of horses. In 1922 she was at last awarded the RCVS diploma on the basis of her studies in 1896–1900, being exempted from the written examination for her war services. The RCVS continued to show no enthusiasm for women veterinarians, who for many years found it difficult to find employment after obtaining its diploma. As late as 1929

Sir John MacFadyean persuaded the Council to pass a resolution that, though it was bound to disregard the sex of students who presented themselves for its diploma, its members would be doing less than their public duty if they did not express their opinion that in competition with men, women in the veterinary profession would always be under severe disadvantages.[4]

DENTISTRY

One profession where the breakthrough came in the 1890s was dentistry. Women had practised as dentists in London as early as the eighteenth century. Several of the most fashionable had come over from Paris and specialised in fitting dentures. When the first dental register was compiled by the General Medical Council under an Act of 1878, 26 women appeared on it. The training of dentists was now regulated in England by the Royal College of Surgeons, which did not admit women to its membership until 1913. When Lillian Murray (later Lindsay) sought to become England's first qualified woman dentist she had no difficulty in being apprenticed to a dentist registered under the Act; but when in 1892 she visited the National Dental Hospital and School in London to apply for enrolment its Dean stood on the doorstep and refused to allow her to enter the building. She was, however, gladly admitted to the Edinburgh Dental Hospital and School by its Dean, William MacLeod, and found the all-male students 'the kindest imaginable'. One of her senior teachers accused her of 'taking the bread out of some poor fellow's mouth' but a tutor, Robert Lindsay, befriended her. When she explained her presence by telling him that she had to earn her living and might as well do so by dentistry as anything else he sternly replied, 'If that is your only reason, give it up.' This shook her. 'From that day I began to realise that the debt I owed to my profession was of equal importance to the debt I owed as a citizen.' In 1895 she qualified with honours for the licentiate in Dental Surgery from the Royal College of Surgeons of Edinburgh, which unlike that of London had no sex restrictions. She thus became the first woman to qualify in dentistry in Britain, though not the first to practise, as women with foreign degrees had already done so. She eventually married her former tutor Robert Lindsay. 'Any use or good I may have been or done', she recollected in old age, 'is entirely due to his example and influence.'[5]

ACCOUNTANTS

It would be hard to think of a profession in which it could be more difficult to justify exclusion of women than that of accountancy. In 1888 the Chairman told the annual meeting of the Society of Accountants and Auditors however,

> it appears that only in regard to one matter have the directors had a difference of opinion which arose through a lady (laughter) claiming to be a qualified public accountant and applying for admission to membership. He was of the opinion that the present time was most inopportune for making such an innovation and his friend Mr Piggott in answer to strong representation on his part had agreed to withdraw his motion in favour of admission of qualified ladies.

The Accountant hoped that 'the gallant Mr Piggott would convert his narrow minded colleagues', but further applications from the lady, Miss Harris Smith, were rejected in 1890 and 1891. That seemed the end of the matter until in 1909 the Society, together with the Institute of Chartered Accountants, sought to persuade Parliament to authorise the two bodies jointly to compile a register of qualified accountants, invoking the precedent of the medical and other professions. They were told by the Board of Trade that no such legislation had a chance of success in favour of organisations which excluded women.

There was still resistance from members who objected that women accountants would work for lower salaries than men and that they would employ women rather than men as their assistants. There was grumbling that admission of women 'would not tend to raise the dignity of the profession'. But the smaller accountancy bodies, which attacked the proposed monopoly of the Society and Institute of Chartered Accountants, were now examining and enrolling women to increase their numbers. At a Special General Meeting of the Society in October 1918, though some members demanded postponement until those serving at the Front returned, admission of women was approved by 55 votes to 13, nominally in recognition of their success in replacing men during the war. In 1919 the Society's Council successfully petitioned a judge in Chancery to agree to extension of its memorandum of association to include women. Miss Harris Smith was thus admitted some 30 years after her first application. Meanwhile she had conducted a successful practice in the City without being a member either of the Society of Accountants and Auditors or of the Institute of Chartered Accountants.[6]

THE LAW

At the beginning of the twentieth century when the women's suffrage movement, both constitutional and militant, was most active, pressure to admit women to the remaining professions mounted. One of the most important was that of the law. By now a few lawyers had begun to urge that women be admitted to their profession. Among them were men who also supported women's suffrage, including Lord Robert Cecil, Lord Wolmer and L. S. Amery. In Parliament the Conservative lawyer J. S. Hills made the question his principal concern. Inside the profession Samuel Garrett, President of the Law Society and the brother of Millicent Garrett Fawcett, was the leading protagonist. The great majority of solicitors opposed the entry of women and made no secret of their reasons in the meetings of the Law Society and in its journal. The Law, they explained, was an adversarial profession in which it would be embarrassing for men to have to confront women. Solicitors, they said, already found it hard enough to make a living, with chartered accountants and house agents nibbling at their business, and they would be ruined if a crowd of women were let loose. Women were quite unsuited to be lawyers because no woman could keep her mouth shut on other people's business. The few solicitors who were in favour of admission of women argued for this on grounds of fairness in what was supposed to be an open profession: they also pointed out that there were considerably more women than men in Britain and that if they were not allowed to earn their living the State might have to support many of those who were unmarried.[7]

Early in 1919 the Government asked the Law Society, representing the solicitors, and the Inns of Court, representing the barristers, for their views on a Bill introduced by Lord Buckmaster in the Lords, proposing that women should be admitted as lawyers. Samuel Garrett when President of the Law Society had found himself with Hills in a minority of two when he tried to persuade it to admit women. When a special meeting was called in March 1919 to reply to the Government's request for the Society's views he moved with passion a motion that the entry of women now be accepted. It was unthinkable, he said, that the women who had played so great a part in the war should be told 'The war is over, now we have done with you.' Lawyers, he told the meeting, were already felt to be wanting in sympathy and out of touch with the public; this feeling would be accentuated if solicitors tried to bar women from their profession just when they had been enfranchised. In any case the meeting could no more stop their entry than a bumble bee could stop a train.

The rejection of Garrett's motion was moved by a London solicitor, G. B. Crook, on the grounds that if women were admitted they would be unfairly privileged because they would be administering laws which, unlike men, they had no obligation to enforce. Warming to his theme, he invoked the lessons of history to prove that no woman had ever been successful in administration. When Queen Mary reigned, he said, she could not keep her axe off the neck of Lady Jane Grey any more than Queen Elizabeth could keep her axe off the neck of Mary, Queen of Scots, or Eve could keep her hands off the apple. Other opponents of the motion reiterated that women were not suited to the profession and that their admission would provide unfair competition to male solicitors returning from the war. One of the latter, however, speaking in uniform, ridiculed this argument, saying that he and his contemporary lawyer soldiers would welcome women as colleagues. Another speaker quoted statistics showing that in medicine only one among twenty-eight doctors was a woman and suggested that among solicitors the proportion was likely to be even less. Garrett's motion was carried by 50 votes to 33 and the Government was informed accordingly. The barristers gave similar advice although, as the Lord Chancellor observed, 'they were not avid' at the prospect of a female invasion of their dinners. At a celebratory dinner given by Hills at which the Lord Chancellor and senior law lords were present, the Law Society was congratulated on yielding gracefully and granting women solicitors the same remuneration as men.[8]

THE CIVIL SERVICE

Perhaps the most important profession of all was the Civil Service, not only because of the number of posts involved but because of the prestige of its administrative class, which, under the broad direction of ministers, implemented the decisions of Parliament and exercised considerable discretion and influence in doing so. Women first came to be employed by the Government somewhat accidentally in 1870 when the Post Office took over private telegraph companies in which they were working as operators. Queen Victoria in 1873 was quoted as being in full sympathy with the struggle for employment of women and suggesting their suitability for this kind of work. Henry Fawcett as Postmaster General from 1880 to 1884 threw open clerkships to women for appointment by competition. The Post Office was the pioneer in this;

other Government departments borrowed women clerks from it to set up their own clerical services staffed by women. Though Fawcett's initiative was inspired by his feminism, his senior officials approved of it on grounds of economy. As one of them saw it, many women clerks would retire to get married and thus not qualify for pensions; they were also less likely to combine for the purpose of extracting higher wages.[9]

These clerks came from the lower middle class. In the 1870s and 1880s however, considerable concern was expressed about the need to find employment for 'women of gentle breeding but small financial resources' such as widows and daughters of men in the Church and State, Army and Navy, letters, the artistic world and the legal and medical professions. These, as an anonymous writer in the *Quarterly Review* pointed out, could hardly take advantage of Fawcett's scheme because

> the daughter of a butler may be quite as good as the daughter of a bishop but we believe there will be a slight awkwardness if they associated and worked at the same desk.... Yet a paralysing weight would be lifted from the mind of many a public servant could a career of usefulness be open to his daughters without loss of social position.

In response to such pleas the Conservative Lord John Manners, who was Postmaster General in 1874–80 and 1885–6, established separate units in the Post Office in which the staff did not need to come in contact with the public, such as the Postal Savings Bank and the clearing house for telegrams and postal orders. 'Ladies' were nominated by him for these posts, though they then had to pass an examination. The Prudential Insurance Office similarly reserved posts for 'ladies'.[10] Whilst the mandarins of the Civil Service accepted women in this kind of capacity, they had a quite different attitude when ministers wished to appoint women to senior posts. When Stansfeld appointed Mrs Nassau Senior as inspector at the Local Government Board in 1873 it was, he said, 'of all his radical policies the thing his officials hated most. Many of them could not endure it.'

The problem of sweated labour was largely one of women working for low pay in unsanitary conditions. When Asquith as Home Secretary in 1892 insisted on appointing women factory inspectors to deal with this his officials protested that their petticoats would get caught up in the machinery. He was defeated by the Prison Commissioners when he wished to appoint a Lady Superintendent in the Prison Service. Almost always the initiative for employment of women in senior posts came from ministers and was resisted by officials. Haldane, who was Secretary for War, told the Royal Commission on the Civil Service in 1912 that there were many higher-division posts which could be filled by women and from which they were excluded by superstition.[11] One exceptional civil servant, Sir Robert Morant, who had a much broader experience outside the service than his colleagues, appointed women as inspectors under the National Insurance Act of 1911. In England they were segregated in a separate division: in Scotland they worked side by side with men and it was possible for a woman inspector to have men working under her.[12]

Faced with the Sex Disqualifications (Removal) Bill of 1919, the senior officials of the Treasury, which controlled admission to the elite Administrative Class of the Civil Service, conducted a skilful policy of quiet obstruction. The Foreign Office, Indian Civil Service and Colonial Service were excluded altogether. For seven years women were unable to compete in the Home Civil Service Examination. The women's most persistent champion in Parliament was the Conservative Sir Samuel Hoare who had served on a Royal Commission on the Civil Service and was experienced in its tactics of resistance to change. In 1919 he told the House 'if we leave it to officials of the Treasury to decide conditions of entry of women they will hedge this around with every kind of restriction . . . this would be a calamity.' Two years later he resumed the attack saying 'The House has declared unanimously for equal treatment of men and women. The Treasury has thwarted it with regulations.' Even Asquith, who had been for so long the greatest obstacle to women's franchise, now told the House that what was good enough for Oxford, which had admitted women to degrees, should be good enough for the Civil Service. When in 1925 women were at last allowed to sit for the Civil Service examination five passed, of whom two went on to reach the top positions of Permanent Secretary. In succeeding years until the 1939 war there were rather few women candidates. They were discouraged by two regulations. First, although the House of Commons passed a resolution in 1920 that women should have equal pay with men, successive Chancellors of the Exchequer were persuaded by their officials that this could not be afforded. The Director of Establishments in the Treasury, Sir Russell Scott, bizarrely told the Tomlin Commission in 1929 that equal pay would bring into the public service women who were too good for their work. Secondly, women who were married were not allowed to enter the Administrative Class and those who married after entering it were obliged to resign.[13]

THE ROYAL ACADEMY OF ARTS

Although the Royal Academy had no official control over the profession of art, election as an academician or associate, with the right to put RA, or ARA after the artist's name, considerably enhanced the fees which could be earned. When King George III founded the Royal Academy in 1768 he included among its first 37 academicians two women, Angelica Kauffmann and Mary Moser, although the Instrument of Foundation limited membership to 'men of fair moral character and of high reputation in their several professions'. The two women voted in the selection of pictures to be hung by sending in marked lists. Mary Moser occasionally attended meetings of the General Assembly.

No other women became members until the twentieth century, though they often exhibited. In 1879, 1880 and 1881 Elizabeth Thompson (later Lady Butler), whose battle scenes were outstandingly popular in the Academy's annual exhibitions, was unsuccessfully nominated for election as an ARA. This led to a general consideration by the Council and Assembly of the position of women. Four resolutions on the subject were voted on in the Assembly in 1880. There was little opposition to women being allowed the titles of RA and ARA but much to their presence at meetings and membership of the Council. Eventually their eligibility for full membership was agreed except for limited access to lectures and schools. One of their strongest advocates was J. E. Millais whose attitude was chivalrous, even sentimental. Only the British, he maintained, had a proper ideal of womankind: the women painted by the Dutch and Italians, he admitted, were magnificent but who, he asked, would want to kiss them? Romantic attitudes, even if sympathetic, could annoy serious women artists. Angelica Kauffmann had complained bitterly that a fellow member in a picture which he exhibited had depicted her in the imagined nude. Despite the resolution passed in 1880 no woman was elected as an ARA until 1922; the first woman RA since the eighteenth century, Dame Laura Knight, was elected in 1936. Not until 1967 did she become the first woman member to be allowed to attend the Society's annual banquet.[14]

There were humbler professions in which women were active from very early times and experienced no problems when a regulatory authority was established. One of these was chiropody. Corn-cutting and foot care developed as a profession in the eighteenth century, the skill often being handed down in a family, and a number of women took to it when so little else was open to them. Thus Charles Dickens' chiropodist, immortalised as Mrs Mowcher in *David Copperfield*, took over the practise from her father. Queen Victoria's chiropodist, Fanny Potter, was trained both by her father and by her mother. When the Society of Chiropodists was formed in 1912–13, 26 of its 128 members were women, two of whom were elected to its Council.[15]

It is the opponents to admission who take up most space in the correspondence and debates recorded in the professional journals, stridently asserting that women are physically weak, emotional and lacking in judgement and that they would lose their precious delicacy and refinement if they were enabled to practise. Some of the women's champions stand out clearly – the lawyer Samuel Garrett, educated in feminism by his dynamic sisters; the young veterinarian Byrne, fired by George Meredith's novels; the chivalrous pharmacist Hampson, indignant at any disparagement of women. Others can only be glimpsed, as they tentatively ask for fair play and are reprimanded by the Chairman for wasting time, and laughed at by their colleagues. It is somewhat curious, in light of outside accusations of 'trade unionism', to find that it was on the whole the older members, close to retirement and with little to lose, who most fiercely resisted the entry of women as being unfair to the younger members, who, though more likely to face female competition, appeared more open minded. Perhaps this was because the former had grown up in professional societies which had started as social gatherings at which their members could talk shop, drink, smoke and relax. Thus their basic, unexpressed motive may have been very similar to that which caused women to be kept out of London's West End Clubs, though it was tactless and improper of the pharmacists' solicitor to make such a comparison.

11 Observations and Reflections

Once this Bill is in law the last fraction of truth about inequality will have gone . . . and the subjection of women, if there be such a thing, will not depend . . . on any action by the law. It will never again be possible to blame the Sovereign State for any position of inequality.

Stanley Baldwin, Prime Minister, speaking in the Commons on the Representation of the People (Equal Franchise) Act (1928)

Among the movements which had important implications for women in the nineteenth century the earliest was that for overall reform of the legal system. There followed the movements for Imperialism, for the Evangelisation of the World in This Generation, and socialism.

Whilst single women from the middle class were preoccupied in the mid-nineteenth century with access to education and to the professions, married women were in revolt against two serious disabilities; it was almost impossible for them to obtain a divorce and they could not own property. Although they and their male feminist friends publicised and sought to remedy these grievances it was eminent lawyers, concerned with tidying up the complicated existing laws and legal procedures affecting them, who were mainly responsible for the necessary legislation being passed.

Before 1857 a man seeking a divorce in England had first to obtain the agreement of an ecclesiastical court on grounds of adultery, then win damages from his wife's lover in a criminal court, and finally have a Bill passed through Parliament. The only grounds on which a woman could obtain a divorce were incest or bigamy. In Scotland, on the other hand, ever since the sixteenth century either a husband or a wife had been able to obtain a divorce in the courts on the grounds of adultery or desertion. Between 1800 and 1857 about 186 men and only 4 women obtained divorces in England. Only the very wealthy could afford the costs. Two elderly former Lord Chancellors, the Tory Lord Lyndhurst and the Whig Lord Brougham, were responsible for the Divorce Act of 1857, which gave considerable relief. Lyndhurst was 85 and Brougham 79 when they urged the House of Lords to change the divorce laws because they were unfair to all except the rich and to all women. They emphasised the absurdity of a situation in which a man divorced in Scotland could be jailed for bigamy if he remarried in England and they asked that the English law should be changed to comply with that of Scotland. The laws of Scotland, Lyndhurst declared, were just and generous to women whilst those in England were cold, unfeeling and barbarous. Both described with indignation how a deserted wife who worked to support herself and her children could legally have her earnings appropriated by her husband and given to his mistress. Lyndhurst stressed the injustice of a law by which if a husband sued her alleged lover the wife was not allowed to give evidence though she had everything at stake.

Brougham was an ardent reformer who had done more than anyone to consolidate the work of civil and ecclesiastical courts. He attacked the divorce laws because they had escaped the general reforms. As a Scot he told the House that the easier and fairer laws in Scotland had done no harm and that 'England, whilst pretending to be a civilised country was in reality living in a system more barbarous than any in the world.' Opposition to change was inspired by High Church Anglicans and by the lawyers in the ecclesiastical courts who would lose their business if divorce procedures were simplified. Samuel Wilberforce, Bishop of Oxford, quoted the Scriptures and St Augustine to show that marriages could not be dissolved. Lyndhurst, old and half blind as he was, retired to study the works of the Early Fathers and returned to refute the Bishop from his own sources. The ecclesiastical lawyers were mollified by being awarded pensions when their courts were abolished. In the House of Commons the Government had a large majority and the Prime Minister, Lord Palmerston, overrode the opposition of Gladstone and other High Churchmen by threatening to keep the House in session all summer until the Bill was passed. Lyndhurst and Brougham failed to change the English law to conform with that of Scotland. The Act of 1857 only gave a wife a right to divorce if adultery were combined with incest, bigamy, cruelty or desertion: a husband, however, could obtain a divorce on grounds of adultery alone. The procedures were greatly simplified and made less expensive by the creation of a Divorce and Matrimonial Causes Court, which consolidated the previous roles of the ecclesiastical and civil courts and of Parliament. This court could deal with custody of children and with maintenance. The double standard was only removed in 1923 when women, like men, were enabled to obtain divorce on grounds of adultery alone.[1]

Until 1870 married women, unlike single women, could not legally own property in England. The authoritative eighteenth-century lawyer

Sir William Blackstone had ruled 'in law husband and wife are one person, and the husband is that person'. A particular disadvantage for an enterprising wife was that because a married woman could not be sued it was very difficult for her to carry on any business. There was an exception for the rich. To meet the needs of the upper classes the Court of Chancery by a very expensive procedure allowed separate property to be settled on a woman under a trustee before she married. Caroline Norton's writings stimulated interest in women's property rights as well as in divorce. A committee of women formed by Barbara Leigh Smith put pressure on Parliament and in 1857 Brougham introduced a Bill to give women the same property rights as men; he likened this to the campaign which he had led for abolition of slavery. Parliament was dissolved before the Bill could progress and it was not until 1868 that a similar Bill was introduced in the Commons, by the Liberal Shaw Lefevre strongly supported by Mill. When opponents quoted Scripture against giving married women property rights they provoked Mill to declare 'every established fact which is too bad to admit any other defence is always presented to us as an injunction of religion'. When it was maintained that to give women independence would encourage immorality, the reformers quoted the examples of Canada, New York State and Massachusetts where similar Bills had been passed without ill effects. A Select Committee to which the Bill was referred confirmed this.

The Conservative Russell Gurney introduced the Bill again in 1869 in the following Parliament, which included many more Radicals. Supported by the Liberal Government, it passed easily through the Commons but was considerably watered down by the Lords. The amended Bill, which was accepted by the Commons hopefully as a first instalment, allowed a married woman to keep her earnings and investments as well as inheriting property. She had to make a special application however to have these registered in her name. Bankers and stockbrokers objected to doing business with women, who still could not be sued for debt.

When the Liberals returned to power in 1880 there was a considerable majority in the Commons for further reform, in which Jacob Bright was a moving spirit. In the Lords the Lord Chancellor Lord Selborne, who had opposed the Act of 1870, came out in favour of the new Bill, which he regarded as a logical consequence of the much broader legal reforms for which he had been responsible under the Judicature Act of 1870. Almost single-handedly in a Government distracted by foreign policy and Irish problems he carried through a sweeping change which, in the Act of 1882, embodied the principles that married women should have the same property rights as single women and that husbands and wives should have separate rights in property. Disputes between spouses could now be settled in a County Court and women could sue and be sued. The practical effects of the Married Women's Property Acts were considerable but the contribution which they made to a new sense of independence among women may have been even more important.[2]

In the late nineteenth and early twentieth century Imperialism reached its highest tide. The leaders of the National League for Opposing Women's Suffrage were two eminent former proconsuls, Lord Cromer, who had ruled Egypt for over 20 years, and Lord Curzon, the dynamic Viceroy of India. They and their associates considered that women ought not to be given the vote because in the end the Empire depended on force, which could only be exercised by men. As has been seen, even after the Act of 1919 women were excluded from the Foreign Office and Indian and Colonial Services. There were some however, such as L. S. Amery, the most prominent Imperialist in Parliament and Government in the next generation, who considered that women should be enfranchised because this would hasten social reforms which were necessary for the stability of the Empire. Across the Atlantic Theodore Roosevelt, who was the closest thing to an Imperialist in American politics, privately favoured women's suffrage for the same kind of reasons and openly endorsed it in his final Presidential campaign.

In general in the Empire the only professional openings for single women were in education and health and as missionaries, though many married women did voluntary social work. Of the first generation of women which emerged from Oxford in the late nineteenth century three-quarters of those who went out to work in the Empire were teachers. Their letters back to their former colleges show them revelling in their liberation from the prospect of purely domestic lives and delighting in the responsibilities of building up new institutions on the model of their Alma Mater. A much broader use of such women might have been made, as may be seen from appointments by two imaginative Imperial disciples of Cromer and Curzon. Both had started life as soldiers. Sir Percy Cox moved from the army to the Indian Political Service and was one of Curzon's protégés as Political Resident in the Persian Gulf. When he became High Commissioner in the new state of Iraq under a British mandate after the First World War he appointed the brilliant Gertrude Bell as Assistant Political Officer, then as Oriental Secretary, because of her knowledge of the tribes; after obtaining a first class degree in Modern History at Oxford in only two years she had spent many years travelling in Arabia. Cox's successor, Sir Henry Dobbs, wrote to the Government in London that she was 'an essential connecting link between the British and Arab races', and that as an official, 'by her sympathy with the infant Iraq state she played a unique part in ensuring close and cordial relations between the High Commission and the Iraq Government'. When the Iraqis became self-governing she stayed on to establish an archaeological museum.[3]

A woman with a very different background, Mary Slessor, was appointed as a Vice Consul and magistrate by Sir Claude MacDonald, who had been Cromer's Military Secretary in Egypt before he was sent out as the first Commissioner to establish an administration in the Oil Rivers Protectorate (now part of Nigeria) in 1891. He recognised that the remote Okoyong tribe in the interior would not take kindly to the presence of young British officials imposing new laws. Mary Slessor was a Scottish Presbyterian missionary who had been toughened by working as a mill hand. She had won the confidence of the Okoyong by her sympathy and understanding of their customs and by negotiating a treaty with hostile coastal tribes to enable the Okoyong to pass freely through their territory to the ports and sell their palm oil. In her new role she presided over her court knitting, usually with an adopted child on her knee, occasionally lambasting a chief with his own ceremonial umbrella if he lied. A woman seldom lost a case in her court. MacDonald's young British officers would visit and learn from her to such an extent that they were accused of 'mariolotry'.[4] MacDonald also did much to help the intrepid anthropologist and trader Mary Kingsley. After her heroic death in the South African war there were many former Imperial officials among those who established the Royal African Society in her memory.

What MacDonald and Cox appreciated was that women with particular knowledge and personality could be valuable in administration not in spite of but because of their femininity. Gertrude Bell could dispose of files in her office efficiently but it was in her house, where she brought British and Iraqis together, that her unique contribution was made. Florence Nightingale after her return from the Crimea had no office; ministers and senior officials would send files to her and come to her home for advice on army sanitary reform and it was de rigueur for outgoing Indian Viceroys to come and be briefed by her. Such procedures would have been inconceivable had she not been a woman. Even Kipling, who was fascinated by how those who ran the Empire went about it, could find a place for the heroine of his story 'William the Conqueror', who proved invaluable when she accompanied her brother on famine relief. 'Life with men who have a great deal of work to do and very little time to do it in', she said, however, 'had taught her the wisdom of effacing, as well as fending for, herself.'[5]

In the nineteenth-century Empire it was possible, if they happened to be interested, for Viceroys and Governors, untrammelled by legislatures and professional associations, to act more rapidly on behalf of women than the Government at home. It has been seen how easily the Governor of Madras was able to arrange the training of women doctors. Similarly J. D. Bethune, Legislative Member of the Governor General's Council, was able to establish a college for girls in Calcutta in 1848 – about the same time as the pioneer Queen's College opened in London and with much less effort. The Governor General, Lord Dalhousie, caused the college to be placed on a firm basis after Bethune's sudden death.[6] Women were enabled to take degrees in Indian universities long before they could do so at Oxford and Cambridge.

Though there was no difficulty in finding women who were eager to serve in the mission field in escape from purposeless lives, until the late nineteenth century the main missionary societies only sent out those who accompanied their members as wives. The pioneer in recruiting single women was Hudson Taylor (1832–1905), a Baptist who in 1864 founded and led what was to become the interdenominational China Inland Mission. Taylor, proclaiming 'in Christ Jesus there is no male or female', was convinced that women were incomparably better than men as evangelists but that married women with domestic responsibilities were inadequate. He recruited single women; the younger they came out the better, he maintained, in order to become fluent in local languages. His missionaries wore Chinese dress. He allowed the women to operate on their own in the interior, confident that they would first convert women and with their help go on to convert men. They were, he said, 'the most powerful army we have at our disposal'. He was strongly attacked in China by missionaries of the established societies who accused him of acting contrary to the Scriptures, of putting young women in physical and moral danger and scandalously lodging them in his own house. Taylor brushed aside the criticism that his approach was unbiblical by explaining that St Paul had only reiterated the headship of husbands, which did not affect single women; his wife dealt sharply with the other charges. Taylor had expected to recruit not so much educated men and women but those from the class that other missionary societies thought beneath their notice. God, he said, has his own universities. To his surprise however, his mission's asceticism made its service immensely attractive to men and women of the upper and middle classes.[7]

In 1892 England was swept by the fervour of the campaign led by the American missionaries Mott and Speer for 'The Evangelisation of the World in This Generation'. In Oxford and Cambridge some of the most brilliant students renounced the prospect of rewarding careers to volunteer for the mission field. By now the mainstream societies were following Taylor's example. Between 1891 and 1900 the Church Missionary Society recruited more women than men. Yet even in 1912 a committee of the Conference of Missionaries noted how women missionaries were 'kept out of authority and responsibility, even the wisest and ablest, subordinate to the most callow tactless young man'. Many of them perhaps found more fulfilment however in helping local women to lead satisfying lives than did the priests and ministers, who were discouraged by their inability to make converts, particularly in Muslim countries. Some women from Oxford and Cambridge escaped male domination by forming the interdenominational Missionary Settlement for University Women in Bombay, whose object was to bring the Gospel to future women leaders of India and which was entirely staffed by women.[8]

In Parliament, as has been seen in Chapter 10, the Labour Party supported women's suffrage, though with a considerable variation of enthusiasm among its members. There were socialists who viewed women as equals and comrades; there were others who saw women as having separate interests which deserved to be represented at the national and local levels. The pure Marxists however, of whom H. M. Hyndman was the most influential, followed the lead of Friedrich Engels in opposing women's suffrage. Engels, in his *Origin of the Family*, published in 1884, argued that in primitive society, in which there was no important private property, there had been sexual freedom for women because paternity was unimportant; women had been valued for child bearing and rearing and for their agricultural work. Later, with civilisation, had come private property and, with it, monogamy because men wished to transmit their possessions to children whom they could be sure were their own. Modern society, Engels asserted, was dependent on the slavery of women. Under Socialism the family unit would disappear, private property would be abolished and the care of children would become a public responsibility. Meanwhile Marxists should not support the struggle for women's rights because this was a middle-class movement. Despite Engels and Marx, the Socialist summer schools, committees and study groups gave women valuable organisational experience and self-confidence which they brought to bear on other causes.[9]

Several times in this study a more liberal attitude towards women in Scotland than in England has been seen. Scotland had fairer divorce laws. The first women were trained as a dentist and as a veterinarian in Scottish institutions when they were barred from those in England.

Early women health inspectors in Scotland were allowed to supervise men when they could not do so in England. Sophia Jex-Blake and her fellow women medical students were shamefully treated, yet it is significant that Edinburgh University was the first in Britain to allow women to matriculate, even though the decision was reversed. In Scotland public opinion did not object to girls being taught by men and there had been a long tradition of co-education in burgh schools. In Ireland too it is notable that Oxford and Cambridge women were able to obtain from Trinity College, Dublin, the formal degree status that they were denied by their own universities. Similarly the King's and Queen's College in Dublin examined them in medicine when no English institution would do so. It has also been seen how an Irish local authority appointed a woman as its veterinary inspector in defiance of Whitehall. Why attitudes to women in the professions in England seem to have been more conservative than those elsewhere in the United Kingdom is a question which seems to deserve more investigation.[10]

The movement for women's equality had a considerable international dimension. Josephine Butler worked closely with societies on the Continent in her campaign for abolition of child prostitution. The leadership in the movement to promote birth control and its publications passed to and fro across the Atlantic. British women went to Paris and the USA to study medicine when they could not do so in Britain. It was the Reverend J. J. May of Syracuse, New York, who inspired Frances Power Cobbe to become a leading suffragist by asking her 'why should you not have a vote? Why should not women be enabled to influence the making of laws in which they have as great an interest as men?'[11] Frequently the successful examples of women obtaining the suffrage and entering universities and the professions in America and continental Europe were quoted in Parliament and elsewhere. Whilst Mill was the recognised pioneer of the movement in Britain and its colonies and was very influential elsewhere, a separate American tradition derived its roots from the discourses on the Rights of Man and the anti-slavery movement, though the comparison between the status of slaves and that of women outraged European feminists. On a number of liberals and radicals who sat at his feet during his long exile in Britain the views of the Italian prophet politician Mazzini (1805–72) made a lifelong impression. Long prejudice, poor education, legal inequality and injustice, he taught, had erected an apparent inferiority in women, who needed to demonstrate that they were capable of moral growth. He became a keen advocate of women's suffrage towards the end of his life.[12]

Campaigners for women's suffrage in Britain were greatly heartened when New Zealand in 1893 became the first country in which women were enabled to vote for the national legislature. The politicians who brought this about were all immigrants from Britain from various backgrounds and parties but each was inspired by a vision of a better society than that which he had left. Virtually all of them had been influenced by reading Mill and many also by working with women in the temperance movement. The Conservative Sir John Hall was a patriarchal sheep farmer who believed that women voters would have a similar interest to that of landlords in social stability. The Jewish Tory Democrat financier Sir Julius Vogel compared the emancipation of women to that of the Jews in Britain, whose emancipation had released a great potential. Radicals and Liberals such as Sir William Fox, Sir Robert Stout, John Ballance and Alfred Saunders were mostly members of the temperance movement who expected that women voters would demand tighter liquor control. Their feminism was part of an egalitarian philosophy and they prided themselves on leading the way for the Mother Country in their social experiments. It was reassuring to British politicians that the women's vote did not alter the balance between the parties in New Zealand and many were encouraged by the social and humanitarian reforms which were a consequence.[13]

As has been seen, of the main newspapers only the *Pall Mall Gazette*, the *Scotsman* and the *Manchester Guardian*, because of the enthusiasm of their editors for women's causes, gave them constant support, whilst *The Times*, which was the most influential in political circles, was generally conservative and hostile. Several of the monthlies would publish sympathetic articles and the weekly *Punch* was friendly in an amused way. Two well-known journalists went so far as to support the Suffragettes. Both H. W. Nevinson and H. N. Brailsford resigned from the staff of the *Daily News*, owned by the Quaker George Cadbury, in 1909 when it supported the forcible feeding of Suffragettes on hunger strike. Nevinson was a crusader by nature who championed the rights of minorities in the Balkans, campaigned against slavery in Portuguese Africa and fraternised with democrats in autocratic Russia and with Home Rulers in India. To him women were one more group deprived of their legitimate rights. He not only wrote about their cause but spoke on WSPU platforms and rode on a splendid horse at the head of their processions, carrying their banner. He was imprisoned for heckling Lloyd George. Ever afterwards he felt that nothing that could happen to him could possibly be quite so difficult, so distasteful and so fertile in ridicule and obloquy as the contest for women's suffrage. The day when the Royal Assent was given to the Bill which conferred the vote on women was, he said, the happiest in his long life.[14] In H. N. Brailsford it was his tutor at Glasgow University, Gilbert Murray, who had first inspired a devotion to women's rights. In his early book *Shelley, Godwin and Their Circle* it was Mary Wollstonecraft who emerged as the heroine, and in writing it he came to believe that men could not be truly free whilst women were enslaved. Like Nevinson he had many other causes, fighting for Greece against the Turks and being prosecuted for providing passports illegally to Russian refugees. He compared the struggle for women's rights to that of national movements in Macedonia, Ireland, Egypt and Russia. Privately he hoped in vain that his support for women's rights might restore an unhappy marriage to a brilliant wife who was a leading figure in the WSPU and whose affection might be won back by this gesture. In this effort he took on the exasperating role of Joint Secretary to Lord Lytton's Conciliation Committee. In this capacity he received considerable help from C. P. Scott, the editor of the *Manchester Guardian*, as a link with Lloyd George.[15]

It is not easy to relate to their backgrounds the attitudes of British decision makers towards the emancipation of women. A considerable number of them had studied at Oxford and Cambridge. Brian Harrison in his book *Separate Spheres* includes a fascinating diagram of the anti-suffrage network at Oxford.[16] A similar diagram might be constructed of the influential pro-suffrage men at Cambridge, including the disciples of Mill and the aristocratic graduates of Trinity College. Yet on the other hand many Oxford men, in particular Gilbert Murray and the Cecils, strongly supported women's suffrage. Cambridge was a few years ahead of Oxford in the establishment of the women's halls, which later were to become colleges. Oxford however, granted degrees to women some 30 years before Cambridge. The divines, classicists and historians who were predominant at Oxford came to favour the emancipation of women partly for economic reasons, seeing that the admittance of their daughters to university courses and eventually to degrees would qualify them to earn their own livings. At the same time the scientists, who were much more numerous at Cambridge, became more articulate in proclaiming the physical and mental inferiority of women. The constitutions of Oxford and Cambridge, which gave every living graduate a voice in policy decisions, caused them to be much more conservative on this issue than the newer universities. Perhaps the intensive study of the civilisation of Ancient Greece and Rome, in which women had few rights, also contributed to this.

Even if the focus is narrowed on to that power house of government, Balliol College, Oxford, between 1870 and 1893, whilst Benjamin Jowett was Master, no clear pattern emerges. Jowett himself sought the advice of Florence Nightingale on many of his schemes, such as the training of future Indian Civil Servants at Oxford, but he was ambivalent about women's suffrage. Among Balliol's graduates Curzon and Asquith were strong opponents of women's suffrage whilst Sir Edward Grey was an equally firm advocate within Asquith's Cabinet. A Balliol man of this generation, J. W. Hills, led the fight within Parliament to open the professions to women. In another Oxford college, Wadham, it was said that two brilliant contemporaries, F. E. Smith (later Lord Birkenhead) and John Simon, tossed a coin to decide which political party each should join, considering that neither could provide sufficient scope for both of them. Smith became the most eloquent opponent of women's suffrage in the Conservative Party whilst Sir John Simon as Home Secretary did much in the final stage to bring it about.

Personal experiences played an important part in shaping attitudes. Many of the women's sympathisers attributed the origin of their feminism to their regard for the abilities of their mothers and to the intellectual fellowship of sisters and wives. In aristocratic circles there was a long tradition of the exercise of political influence by wives through their husbands: the extension of the suffrage to their male servants whilst it was withheld from their brilliant female relatives was a powerful force in the conversion of aristocrats to women's suffrage. Some employers supported emancipation of women because it would produce cheap semi-professional labour, some lawyers because it would tidy up the law, some middle-class men because their wives did not wish to be treated by male doctors. Yet perhaps the most numerous supporters were those who, without exploring the issues deeply, had an uneasy feeling that the disabilities of women were unfair.

In the end it was Parliament which decided the rate of progress of women's emancipation. The tempo varied, often depending on how far the interests of members and their prospects of re-election were involved. Its response to Josephine Butler's campaign for abolition of the Contagious Diseases Bill was slow because women of its own class were not affected. On the other hand, when threatened by exposure of the behaviour of its members on the eve of a General Election it rapidly capitulated and passed legislation to protect young girls from prostitution, in the wave of indignation which followed Stead's lurid revelations. It agreed, with little opposition, that the new universities should admit women to degrees whilst it would not or could not insist that Oxford and Cambridge, of which so many of its members were graduates, should do the same. On the whole it took a more liberal attitude to admission of women to the professions than did the leaders of most of those professions. It was prepared to override the doctors' objections to admitting women to their ranks and to the professionalisation of midwives. The accountants were told that their associations could not be given regulatory powers unless they admitted women. The lawyers were treated more tenderly, having a powerful position in both Houses. The Sir Humphreys of the Civil Service managed to delay the penetration of all but a few women into their senior ranks but Members of Parliament of all parties harried them on the question.

Parliament reformed the divorce laws quite speedily when their unfairness, expense and delays were exposed, but men as well as women had a stake in this. The right of married women to own property could not be fully established while Conservative governments were in office but on most women's issues there were some Conservatives, such as Lord Shaftesbury and Russell Gurney, who joined with Liberals in effecting reforms.

Women were admitted into universities and the professions in Britain more slowly than in the USA, partly because in America individual States could go ahead at their own pace. On one issue however, the stolidity of British legislators made them more sympathetic to women than their American counterparts. Parliament was not intimidated by the Church of England into passing laws to curtail dissemination of information on contraception, whilst Congress and some State legislatures did so under pressure from the Roman Catholic Church. There were Members of Parliament who championed all or most women's causes, including Mill, Stansfeld, Jacob Bright and Russell Gurney. Others, such as Curzon and James Bryce, whilst opposing the suffrage, strongly supported higher education for women and their right to university degrees.

It was on the suffrage question that Parliament caused most exasperation by subordinating principles to self interest. For many years, although a majority of its members favoured reform when there was a Liberal Government, implementation was again and again delayed because of apprehension as to how the various measures proposed would affect party interests. Indignant at witnessing from the Gallery the Parliamentary devices used to prevent progress, the Suffragettes embarked on a campaign of violence to which even feminist Members felt bound to respond by making no concessions; thus there was a vicious circle.

Yet, apart from Australia and New Zealand and the special cases of Finland and Norway, Britain was eventually one of the earliest countries to give women the Parliamentary vote, at about the same time as the USA and the European countries in which Protestants were in a majority. Whether because the Latin countries were Roman Catholic or because they were strongly influenced by legal and social traditions of the Roman Empire, in France, Italy and Spain women's suffrage only came after the Second World War. Although France had been one of the earliest countries in which women's rights had received serious consideration, in the late nineteenth century Republicans and Radicals became fearful that women if enfranchised would vote for Church-influenced parties.

Stanley Baldwin's claim in 1928, quoted at the head of this chapter, that no further action by the law would ever again be required to redress the inequalities of women proved to be too complacent: some forty years later Parliament felt it necessary to pass the Sex Discrimination Act. Inequalities in employment, pay and pensions were prohibited and wives were now enabled to be taxed separately from their husbands. Women came to be admitted to men's colleges at Oxford and Cambridge which had so long been the nurseries of the Establishment, and even to the London Clubs which were the haunts of the decision makers. More broadly the provision of free contraception and of nurseries for children of women workers were important factors in women's further liberation. In 1997 positive discrimination in selection of candidates for the first time gave women a substantial representation in Parliament when Labour came into power. The part which men played in this second wave of women's emancipation lies outside the scope of this study. Necessary as it was, it did not require the courage needed by those of their grandfathers who had incurred ridicule and unpopularity and even sacrificed careers in the cause.

Notes

Notes to Chapter 1: Introduction

1. C. Dickens, *Bleak House* (London, 1896 edition) Ch. xxx. *Household Words*, 8 November 1851, p. 145. P. Collins, *Dickens and Education* (London, 1963) pp. 124ff. R. Symonds, *Far above Rubies* (Leominster, 1993) p. 109.
2. A. Trollope, *Is He Popinjoy*? World Classics edition (Oxford, 1986) Ch. xvii and Appendix ix. V. Glendinning, *Trollope* (London, 1992) pp. 325, 480 and passim.
3. A. Woods, *George Meredith: Champion of Women and of Progressive Education* (Oxford, 1937) pp. vii, 19, 22, 25, 79.
4. M. Millgate, *Thomas Hardy: A Biography* (Oxford, 1982) pp. 192, 357, 369. R. Sumner, *Thomas Hardy: Psychological Novelist* (London, 1981) p. 190.
5. G. B. Shaw, *The Quintessence of Ibsenism* (London, 1896) pp. 41ff. R. Weintraub (ed.), *Fabian Feminists: Bernard Shaw and Women* (University of Pennsylvania, 1977) pp. 2–10. S. Strauss, *Traitors to the Masculine Cause* (London, 1982) pp. 156–60. G. B. Shaw, *Major Barbara* (London, 1905); *Mrs Warren's Profession* (London, 1925); *Press Cuttings* (London, 1926).
6. H. G. Wells, *Ann Veronica* (London, 1909).
7. J. A. Froude, *Life of Carlyle*, ed. S. Clubbe (London, 1979) p. 10. T. Carlyle, *French Revolution* (London, 1902 edn) vol. ii, p. 58; vol. iii, p. 245.
8. J. Ruskin, *Works*, ed. E. T. Cook and A. Wedderburn (1905) vol. 34, p. 499; vol. 18, p. 123; vol. 23, p. 332; vol. 27, p. 619.
9. T. H. Huxley, *Life and Letters* (London, 1900) vol. i, pp. 38, 212, 417; *Lay Sermons* (London, 1870) p. 27 (originally published 1865 in *The Reader*).
10. A. Tennyson, *Works* (London, 1891) pp. 165ff. J. Killham, *Tennyson and 'The Princess'* (London, 1958) passim. W. C. Gordon, *Social Ideals of Tennyson* (London, 1906) pp. 74ff. Stopford Brooke, *Tennyson* (London, 1910) p. 164. M. Thorn, *Tennyson* (London, 1992) p. 20.
11. Stopford Brooke, *The Poetry of Robert Browning* (London, 1902) chs xiii and xiv.

Notes to Chapter 2: John Stuart Mill Raises the Standard

1. J. S. Mill, *Dissertations and Discussions* (London, 1859) vol. ii, p. 411, 'The Enfranchisement of Women'.
2. F. A. Hayek, *J. S. Mill and Harriet Taylor* (London, 1951) p. 26.
3. M. St J. Packe, *Life of J. S. Mill* (London, 1954) p. 110.
4. Thomas Carlyle, *Reminiscences* (London, 1881) vol. ii, p. 177. Packe, *Life of J. S. Mill*, p. 180.

5. Packe, *Life of J. S. Mill*, p. 403.
6. Ibid., p. 180.
7. Hayek, *J. S. Mill and Harriet Taylor*, p. 17.
8. Packe, *Life of J. S. Mill*, p. 56.
9. Ibid., p. 125.
10. Ibid., p. 317.
11. Mill, *Dissertations*, vol. II, pp. 411ff.
12. Hansard, *Parliamentary Debates*, 20 June 1866, cols 817–29, speech by J. S. Mill.
13. Ibid., cols 830ff.
14. Hansard, *Parliamentary Debates*, 27 April 1866, col. 99, speech by B. Disraeli.
15. Packe, *Life of J. S. Mill*, p. 501.
16. J. S. Mill, *Autobiography* (1924 edition) p. 185.
17. J. S. Mill, *The Subjection of Women* (London, 1878 edition).
18. Packe, *Life of J. S. Mill*, p. 495. A. Bain, *John Stuart Mill: A Criticism* (London, 1882) pp. 130–2. Frederic Harrison, *Tennyson, Ruskin, Mill* (London, 1899) p. 310.
19. T. Martin, *Queen Victoria as I Knew Her* (London, 1908) p. 69.
20. *The Times*, London, 10 May 1873.
21. J. Morley, *Life of W. E. Gladstone* (London, 1904) vol. 2, p. 543. W. D. Christie, *J. S. Mill and Mr Abraham Hayward Q. C.* (London, 1873).

Notes to Chapter 3: The Allies of Josephine Butler

1. Josephine Butler, *Personal Reminiscences of a Great Crusade* (London, 1896) p. 20.
2. See R. Symonds, *Alternative Saints: The Post-Reformation People Commemorated by the Church of England* (London, 1988) p. 8.
3. Josephine Butler, *Recollections of George Butler* (London, 1892) pp. 19, 64, 65.
4. Josephine Butler, *An Autobiographical Memoir* (Bristol, 1904) p. 15.
5. Butler, *Recollections*, p. 57; E. Moberly Bell, *Josephine Butler* (London, 1962) p. 28.
6. Josephine Butler (ed.), *Women's Work and Women's Culture* (London 1869) Introduction.
7. Ibid., pp. 49ff.
8. Butler, *Recollections*, pp. 219–20.
9. Ibid., p. 230.
10. Ibid., pp. 240, 245.
11. Ibid., p. 248.
12. Ibid., p. 257.
13. Ibid., p. 384.
14. Ibid., pp. 483, 215.
15. W. T. Stead, *Josephine Butler: A Life Sketch*; Josephine Butler, *Autobiographical Memoir* (Bristol, 1911) p. 34.
16. Butler, *Autobiographical Memoir*, p. 35.

17. H. Temperley, *British Antislavery* (London, 1972) pp. 88ff.
18. J. L. and Barbara Hammond, *James Stansfeld: A Victorian Champion of Sex Equality* (London, 1932) p. 286.
19. *Dictionary of National Biography* (*DNB*), vol. I, p. 654, W. T. Ashurst.
20. Anon., *Memorable Unitarians* (London, 1906) p. 310.
21. *The Times*, London, 18 February 1898.
22. Hammond, *James Stansfeld*, p. 41.
23. Ibid., p. 189.
24. G. Petrie, *A Singular Iniquity: The Campaigns of Josephine Butler* (London, 1971) p. 149.
25. Hammond, *James Stansfeld*, p. 191.
26. Ibid., p. 203.
27. Hansard, *House of Commons Debates*, 23 Jan 1875, cols 409–14.
28. *Report of the Select Committee of the House of Commons on the Contagious Diseases Acts*, Parliamentary Papers 1881, IX. The minority report is included.
29. Hansard, *House of Commons Debates*, 20 April 1883, cols 764ff.
30. Ibid., cols 823, 840.
31. J. A. Spender, *Life of Sir H. Campbell Bannerman* (1923) vol. 1, p. 105.
32. Hansard, *House of Commons Debates*, 16 March 1886, cols 981ff.
33. Ibid., p. 287.
34. Ibid., p. 288.
35. Ibid., p. 292.
36. Ibid., p. vii; M. Fawcett and E. M. Turner, *Josephine Butler* (London, 1927) Preface.
37. F. Whyte, *Life of W. T. Stead* (London, 1925) vol. 1, p. 100.
38. *Report of Select Committee of House of Lords Relating to Protection of Young Girls*, Parliamentary Papers, Sessional Papers, 1882, vol. VII.
39. Whyte, *Life of W. T. Stead*, vol. I, p. 160.
40. Hansard, *Parliamentary Debates*, 22 May 1885, cols 1175ff.
41. All the quotations in the previous three paragraphs are from the *Pall Mall Gazette*, London, 6, 7, 8, 9 July and 11 November 1885.
42. Hansard, *Parliamentary Debates*, 6 August 1885, cols 1409ff.
43. Petrie, *A Singular Iniquity*, p. 256.
44. Moral Reform Union, London 1885, *Speech by Mr W. T. Stead at the Central Criminal Court.*
45. W. T. Stead, *My First Imprisonment* (London, 1886); and Whyte, *Life of W. T. Stead*, vol. 1, p. 207.
46. Whyte, *Life of W. T. Stead*, p. 314.

Notes to Chapter 4: Emancipation through Birth Control

1. P. Fryer, *The Birth Controllers* (London, 1965) pp. 31ff. Richard Symonds and Michael Carder, *The United Nations and the Population Question* (London, 1973) p. 22.
2. N. Himes, *Place on Population Control* (London, 1930). Fryer, *The Birth Controllers*, pp. 64ff.

3. Fryer, *The Birth Controllers*, p. 76. G. A. Aldred, *Richard Carlile, Agitator* (London, 1923).
4. R. W. Leopold, *Robert Dale Owen: Radical Reformer* (London, 1940). Fryer, *The Birth Controllers*, p. 92.
5. *New England Quarterly*, vol. VI (1933) p. 470. Fryer, *The Birth Controllers*, p. 99.
6. G. Drysdale, *Physical, Sexual and Natural Religion* (later called *Elements of Social Science*) (London, 1855). Fryer, *The Birth Controllers*, p. 111.
7. A. H. Nethercot, *The First Five Lives of Annie Besant* (Chicago, 1960) pp. 107ff.
8. Fryer, *op.cit.*, p. 169.
9. Bertrand and Patricia Russell, *Amberley Papers* (1937) vol. II, p. 171.
10. Fryer, *The Birth Controllers*, p. 116.
11. Himes, *Place on Population Control*, p. 256. Dr H. A. Allbutt, *The Wives' Handbook* (London, 1883).
12. Fryer, *The Birth Controllers*, p. 130.
13. *Nineteenth Century and After* (1900) vol. 59, p. 80. Fryer, *The Birth Controllers*, p. 181.
14. 'The Declining Birthrate', Report of the Commission of Enquiry set up by the National Council of Public Morals (London, 1916) p. 242.
15. Ibid., p. 273.
16. Ibid., p. 371. S. Szrecer, *Fertility and Gender in Britain, 1860–1940* (Cambridge, 1996) pp. 367ff.
17. A. MacLaren, *Birth Control in 19th Century England* (London, 1978) p. 56. *DNB* article, Richard Carlile.
18. *Fortnightly Review*, March 1888, p. 119.
19. *The Times*, 21 October 1905, p. 14, col. 5.
20. *The Six Lambeth Conferences, 1867–1920* (London, 1929), 1908 Conference, pp. 310, 327, 342–99.
21. Ibid., 1920 Conference, p. 44, 112.
22. F. Watson, *Dawson of Penn* (London, 1951) p. 55 and passim.
23. Ibid., p. 159.
24. Lord Dawson of Penn, *Love, Marriage and Birth Control, Speech Delivered at the Church Congress in Birmingham* (London, 1922).
25. Fryer, *The Birth Controllers*, p. 245.
26. Dawson, *Love, Marriage and Birth Control*, Introduction.
27. *The Times*, 11 January 1922, p. 7.
28. *Lambeth Conferences, (1867–1930)* (London, 1968) p. 164.
29. G. L. Prestige, *Life of Charles Gore* (London, 1935) pp. 515, 517.
30. Bishop A. A. David and Bishop M. B. Furse, *Marriage and Birth Control* (London, 1932) p. 30.
31. Hansard, *House of Lord's Debates*, vol. 90, 13 February 1934, col. 818.
32. David and Furse, *Marriage and Birth Control*, pp. 13, 27.
33. *Lambeth Conference – What the Bishops said about Marriage* (London: SPCK, 1968).
34. Fryer, *The Birth Controllers*, p. 249.
35. Himes, *Place on Population Control*, p. 286.
36. Fryer, *The Birth Controllers*, p. 117.
37. Ibid., p. 197.

38. D. M. Kennedy, *Birth Control in America: The Career of Margaret Sanger* (Yale, 1970) p. 80.
39. Fryer, *The Birth Controllers*, pp. 216, 219.
40. Kennedy, *Birth Control in America*, pp. 218ff.
41. Ibid., pp. 98, 193.
42. June Rose, *Marie Stopes and the Sexual Revolution* (London, 1992) pp. 210ff.

Notes to Chapter 5: Gandhi and Liberation through the Freedom Movement

1. *Collected Works of Mahatma Gandhi* (hereafter *CWMG*), vol. XXXV (1927) p. 44, Public Meeting, Paganeri.
2. Mira Behn, *The Spirit's Pilgrimage* (London, 1960) p. 200.
3. *CWMG*, vol. XIV (1907) p. 26, speech at Gujerati Educational Conference. G. Forbes, 'The Politics of Respectability', in D. A. Low (ed.), *The Indian National Congress* (Bombay: Oxford University Press, 1988) p. 64.
4. B. R. Nanda, *Gandhi and His Critics* (Delhi, 1985) p. 39.
5. *CWMG*, vol. XIV (1917) p. 14, speech to Gujerati Political Conference.
6. *CWMG*, vol. XXII (1922) p. 181, article in *Navajivan*.
7. M. Kishwar, 'Gandhi on Women', *Economic and Political Weekly* (Bombay) vol. 20, no. 40 (1985).
8. G. Minault, *The Extended Family* (Delhi, 1981) p. 185. Also *CWMG*, vol. XL (1929) p. 417, article in *Young India*.
9. J. Nehru, *Discovery of India* (London, 1945) p. 23.
10. *CWMG*, vol. XXXIII (1927) p. 333, letter to Ashram Women.
11. V. L. Pandit, *The Scope of Happiness* (London, 1979) pp. 110, 171.
12. *CWMG*, vol. XVII (1920) p. 49, speech at Meeting of Mill Hands, Ahmedabad.
13. *CWMG*, vol. XLVIII (1931) p. 79, interview with *Daily Herald*.
14. P. Joshi, *Gandhi on Women* (New Delhi, 1988) p. 317.
15. *CWMG*, vol. XXIV (1924) p. 498, speech at National Education Conference; and vol. XXXIV (1927) p. 384, letter to Anandibai.
16. *CWMG*, vol. XVII (1920) p. 536, speech to students at Satyagraha Ashram; and vol. XXVII (1925) p. 151, article in *Navajivan*.
17. *CWMG*, vol. LVIII (1931) p. 311, speech at meeting of Women's Indian Council, London.
18. *CWMG*, vol. LXXV (1941) p. 155, 'The Constructive Programme'.
19. *CWMG*, vol. LXVIII (1939) p. 312, interview with representatives of municipalities.
20. *CWMG*, vol. LXVIII (1939) p. 312, interview with representatives of municipalities; vol. XXII (1922) p. 188, 'My Notes'.
21. S. Natarajan, *A Century of Social Reform in India* (Bombay, 1959) p. 147.
22. *CWMG*, vol. LXII (1935) p. 156, 'Interview to Margaret Sanger'.
23. Foreword by Sarojini Naidu, in H. Polak et al., *Mahatma Gandhi* (London, 1948).
24. V. S. Naravani, *Sarojini Naidu* (New Delhi, 1980), p. 66 and passim.

25. Mira Behn, *The Spirit's Pilgrimage* (London, 1960). Ved Mehta, *Mahatma Gandhi and his Disciples* (London, 1977).
26. M. K. Gandhi, *The Story of My Experiments with Truth* (Ahmedabad, 1927); and E. H. Erikson, *Gandhi's Truth* (London, 1970) p. 121.
27. *CWMG*, vol. XLI (1929) p. 405, article 'Service to Women'.
28. *CWMG*, vol. XXXIX (1929) p. 415, speech at DJS College, Karachi.
29. *CWMG*, vol. XL (1929) p. 417, article in *Young India*.
30. *CWMG*, vol. LXXVIII (1941) p. 236, discussion with Hindustani Talimi Sangh.

Notes to Chapter 6: Education

1. Sydney Smith, *Selected Writings* (London, 1957) pp. 271–81; Gillian Sutherland, 'The Plainest Principles of Justice', in F. M. C. Thompson (ed.), *The University of London and the World of Learning* (London, 1990).
2. R. G. Grylls, *Queen's College, 1848–1948* (London, 1948) p. vii.
3. E. Kaye, *History of Queen's College* (London, 1972) p. 11.
4. F. McLain, *F. D. Maurice: A Study* (Cowley, 1982) p. 32.
5. F. Maurice, *Life of F. D. Maurice* (London, 1884) vol. I, pp. 57ff.
6. Ibid., vol. II, p. 32.
7. Maurice's inaugural address is contained in A. Tweedle, *The First College for Women* (London, ND).
8. Grylls, *Queen's College*, p. 28. Kaye, *History of Queen's College*, p. 182.
9. McClain, *F. D. Maurice*, p. 47.
10. Ibid., p. 52.
11. Maurice, *Life*, p. 641.
12. Lady Eastlake, *Mrs Grote: A Sketch* (London, 1880) p. 43.
13. George Grote, *Minor Works* (London, 1873) p. 162. N. B. Harte, *The Admission of Women to University College, London*, Centenary Lecture (London, 1979) p. 5.
14. Harte, *The Admission of Women*, p. 9 and passim. H. S. Solly, *Life of Professor Henry Morley* (London, 1898) p. 308 and passim.
15. A. Harrison-Barbet, *Thomas Holloway: Victorian Philanthropist* (Royal Holloway College, 1994) p. 58.
16. Ibid., pp. 50, 52.
17. Ibid., pp. 54ff.
18. C. Bingham, *History of Royal Holloway College* (London, 1987) p. 22.
19. Harrison-Barbet, *Thomas Holloway*, p. 90.
20. Bingham, *History of Royal Holloway College*, p. 59.
21. Ibid., pp. 73, 91.
22. J. Bryce, *Studies in Contemporary Biography* (London, 1903) p. 328.
23. R. McWilliams-Tullberg, *Women at Cambridge* (London, 1975) pp. 38–9. D. Bennett, *Emily Davies and the Liberation of Women* (London, 1990) passim.
24. R. Skidelsky, *Interests and Obsessions* (London, 1993) pp. 3ff.
25. A. Sidgwick and E. M. Sidgwick, *Henry Sidgwick: A Memoir* (London, 1906) p. 189.

26. McWilliams-Tullberg, *Women at Cambridge*, p. 49.
27. E. Sidgwick, *Mrs Henry Sidgwick* (London, 1938) p. 40.
28. A. and E. M. Sidgwick, *Henry Sidgwick: A Memoir*, p. 59.
29. Ibid., pp. 226, 268, 250. *University Degrees for Women*, Report of Conference on Royal Holloway College, December 1897, p. 33.
30. E. Sidgwick, *Mrs Henry Sidgwick*, pp. 52, 119.
31. Ibid., p. 123. A. and E. M. Sidgwick, *Henry Sidgwick: A Memoir*, pp. 544ff.
32. E. Sidgwick, *Mrs Henry Sidgwick*, p. 252.
33. A. and E. M. Sidgwick, *Henry Sidgwick: A Memoir*, p. 199.
34. E. Sidgwick, *Mrs Henry Sidgwick*, p. 153.
35. McWilliams-Tullberg, *Women at Cambridge*, pp. 155ff.
36. E. Wordsworth, *Glimpses of the Past* (London, 1913). P. Adams, *Somerville for Women* (Oxford, 1996). G. Stephenson, *Edward Stuart Talbot* (London, 1936).
37. S. Fletcher, *Feminists and Bureaucrats* (Cambridge, 1980) p. 40 and passim.
38. J. A. K. Thompson and A. Toynbee, *Essays in Honour of Gilbert Murray* (London, 1936) p. 49.

Notes to Chapter 7: Medicine

1. E. Blackwell, *Pioneer Work for Women* (London, 1914 edition) p. 21.
2. Ibid., p. 28.
3. Ibid., p. 53.
4. Ibid., p. 233.
5. Blackwell, *Pioneer Work for Women*, p. 145.
6. Ibid., p. 196. M. S. Fancourt, *They Dared to be Doctors* (London, 1965) p. 126.
7. J. Manton, *Elizabeth Garrett Anderson* (London, 1965) p. 351.
8. Ibid., pp. 112, 114.
9. Ibid., p. 120.
10. Ibid., p. 130.
11. *British Medical Journal*, 22 November 1862.
12. Manton, *Elizabeth Garrett Anderson*, p. 147.
13. Ibid., p. 163.
14. Ibid., p. 255.
15. Ibid., p. 291.
16. L. G. Anderson, *Elizabeth Garrett Anderson* (London, 1939) p. 182. Manton, *Elizabeth Garrett Anderson*, p. 147.
17. S. Roberts, *Sophia Jex-Blake* (London, 1993) pp. 22, 61.
18. M. Todd, *The Life of Sophia Jex-Blake* (London, 1918) p. 299.
19. Roberts, *Sophia Jex-Blake*, pp. 138–9; J91–108.
20. C. Reade, *A Woman Hater* (London, 1877) vol. III, pp. 273ff, and passim. M. Elwin, *Charles Reade* (London, 1931).
21. Elwin, *Charles Reade*, p. 321.
22. Hansard, *House of Commons Debates*, 12 June 1874, cols 1526ff.
23. Hansard, *House of Commons Debates*, 3 March 1875, cols 1123ff.

24. Hansard, *House of Commons Debates*, 5 July 1876, cols 1003ff.
25. J. B. Atlay, *Sir Henry Wentworth Acland* (London, 1903).
26. H. W. Acland, 'Medical Education for Women', letter of 25 April 1870 to *The Times*, printed privately (Bodlelan Library, Oxford).
27. The relevant debates and resolutions are in the minutes of the Medical Council, London, 1875 and 1876.
28. Roberts, *Sophia Jex-Blake*, pp. 143–5; Catriona Blake, *The Charge of the Parasols: Women's Entry to the Medical Profession* (London, 1990) p. 188 and passim.
29. Roberts, *Sophia Jex-Blake*, p. 159.
30. M. Scharlieb, *Reminiscences* (London, 1924) p. 90.
31. *Life of Sir Robert Christison*, edited by his sons (1886) vol. 2, p. 43.
32. R. B. Fisher, *Joseph Lister* (London, 1977) p. 189 and passim.
33. J. Duns, *Memoir of Sir James Simpson* (Edinburgh, 1873).
34. *Life of Sir R. Christison*, op. cit., vol. 11, pp. 43–50.
35. Sir James Paget, *Memoirs and Letters* (London, 1903) p. 298.
36. Blake, *Charge of the Parasols*, p. 172 and passim.
37. D. Masson, *Lecture to Edinburgh Ladies' Education Association* (Edinburgh, 1868).
38. H. E. Graham, *Literary and Historical Essays* (London, 1908) p. 219.
39. G. Travers, *Life of Sophia Jex-Blake* (London, 1908) p. 303.
40. J. Stansfeld, 'Medical Women', in *Nineteenth Century*, July 1877. Catriona Blake's, *The Charge of the Parasols* is a useful overall account of the campaign.
41. J. Dennison, *Midwives and Medical Men* (London, 1977) is the source for this section.

Notes to Chapter 8: Religion

1. E. Isichei, *Victorian Quakers* (Oxford, 1970) pp. 102–10 and passim. J. Butler, *Reminiscences of a Great Crusade* (London, 1896) pp. 20, 60–3. H. Ausobel, *John Bright: Victorian Reformer* (London, 1966) p. 195. K. Robbins, *John Bright* (London, 1979) p. 214. *The Friend*, 1914, accounts of sessions of Meeting for Sufferings, Friends House, London.
2. R. F. Watts, 'The Unitarian Contribution to Female Education in the 19th Century', unpublished thesis, Manchester College, Oxford (1981) p. 46. K. Gleadle, *Early Feminists, Radical Utilitarians and the Emergence of Women's Rights* (London, 1995) p. 28.
3. Gleadle, *Early Feminists*, p. 28.
4. S. K. Ratcliffe, *The Story of South Place* (London, 1955) p. 18. William Johnson Fox, *Collected Works*, vol. VI (London, 1867) p. 180. Ibid., vol. V (1866) p. 283.
5. Ibid., vol. VI (1867) p. 164.
6. M. J. Shaen, *William Shaen* (London, 1912) pp. 8, 15, 56 and 87.
7. Bramwell Booth, *Echoes and Memories* (London, 1926) p. 172.
8. H. Begbie, *Life of William Booth* (London, 1920) vol. 1, p. 27.

9. Bramwell Booth, *Echoes and Memories*, p. 166. F. de L. Booth-Tucker, *Life of Catherine Booth* (London, 1893) vol. 1, p. 359.
10. Anon., *Twenty-One Years of the Salvation Army* (London: Salvation Army, 1887) p. 105.
11. *Orders and Regulations for Field Officers* (Salvation Army, 1886) Preface. Ibid. (1900 edition) p. 294.
12. St John Irvine, *God's Soldier* (London, 1934) vol. I, p. 608; vol. II, p. 962.
13. Major J. Fairbank, in *The Officer*, October 1988. Mrs Commissioner Hodder, in *The War Cry*, 23 October 1993.
14. D. M. Paton, *'R.O.': The Life and Times of Bishop Ronald Hall of Hong Kong* (London, 1985) pp. xiii and 120.
15. Ibid., pp. xii and 15.
16. Ibid., pp. 42, 73.
17. Ibid., p. 125.
18. Ibid., pp. 127–30.
19. Ibid., p. 133.
20. Ibid., p. 136.
21. Ibid., p. 140.
22. Florence Tim Oi Li, *Much Beloved Daughter* (London, 1985) pp. 45ff.
23. Ibid., p. 114.
24. Paton, *'R.O.'*, p. 141.
25. J. G. H. Baker, *Bishop Speaking* (Hong Kong, 1981). Joyce Bennett, *Hasten Slowly* (Chichester, 1991) pp. 13, 15.
26. R. Strachey, *Millicent Garrett Fawcett* (London, 1931) p. 88.

Notes to Chapter 9: Parliament and Suffrage

1. Leslie Stephen, *Henry Fawcett* (London, 1885). Hansard, *House of Commons Debates*, 20 March 1867, col. 835; 30 April 1872, col. 1239; 26 April 1876, col. 1711. H. Fawcett, *Speeches* (London, 1873) pp. 161ff.
2. Hansard, *House of Commons Debates*, 6 June 1877, col. 1114; 7 March 1879, col. 405; 6 July 1883, col. 1113; 18 February 1886, col. 689; 27 April 1892, col. 1454; 3 February 1897, col. 1330. L. Courtney, *Cornish Granite* (London, 1925).
3. Hansard, *House of Commons Debates*, 4 May 1870, col. 216; 2 March 1906, col. 1448. Sir Charles Dilke, *Woman Suffrage and Electoral Reform* (London, ND [? 1910]. D. Nicholls, *The Lost Prime Minister – A Life of Sir Charles Dilke* (London, 1995).
4. Jacob Bright, *Speeches* (London, 1885) pp. 210 and 332. Hansard, *House of Commons Debates*, 30 April 1873, col. 1174; 26 April 1876, col. 1687; 6 June 1877, col. 1706; 6 July 1883, col. 454. E. Isichei, *Victorian Quakers* (Oxford, 1970) p. 109 and passim. *Vanity Fair*, 5 May 1877; *Manchester Faces and Places*, 1889, p. 181. Sylvia Pankhurst, *The Suffragette Movement* (London, 1931) passim. *DNB*, William Woodall.
5. Hansard, *House of Commons Debates*, 7 April 1884, col. 454; 12 June 1884, col. 106.
6. J. Lewes, *Before the Suffrage was Won* (London, 1987) p. 444.

7. Hansard, *House of Commons Debates*, 27 April 1892, col. 1501. H. A. L. Fisher, *James Bryce* (London, 1927) vol. I, p. 188; vol. II, p. 230.
8. Hansard, *House of Commons Debates*, 8 March 1907, col. 1110.
9. M. Thomson, *David Lloyd George* (London, ND) pp. 206, 221. J. Grigg, *Lloyd George* (London, 1928) p. 298.
10. Hansard, *House of Commons Debates*, 6 May 1913, col. 1938.
11. F. Maurice, *Haldane* (London, 1937) p. 56. Hansard, *House of Commons Debates*, 11 July 1910, col. 82.
12. J. Scheer, *George Lansbury* (Manchester, 1990) pp. 84, 86, 119, 124, 197.
13. J. Keir Hardie, *From Serfdom to Socialism* (London, 1907) pp. 61ff. W. Stewart, *J. Keir Hardie* (London, 1921). Hansard, *House of Commons Debates*, 24 January 1913, col. 1086.
14. Hansard, *House of Commons Debates*, 8 March 1907, col. 1133; 28 March 1912, col. 205; 5 May 1913, col. 1705.
15. Ibid., 27 April 1866, col. 99.
16. G. W. F. Russell, *Portraits of the Seventies* (London, 1916) p. 232. K. Rose, *The Later Cecils*, (London, 1975) p. 27.
17. Hansard, *House of Commons Debates*, 6 May 1913, col. 1118; 28 March 1912, col. 664; 19 June 1917, col. 1735.
18. Hansard, *House of Commons Debates*, 27 April 1892, col. 1524; K. Young, *Arthur James Balfour* (London, 1963) p. 320.
19. Hansard, *House of Commons Debates*, 24 June 1913, col. 882.
20. Ibid., 27 April 1892, col. 1505.
21. Ibid., 12 June 1884, col. 93.
22. Unpublished memoir by C. M. Woodhouse, Lytton Archives, Knebworth.
23. Hansard, *House of Commons Debates*, 16 March 1904, col. 1361; 25 April 1906, col. 1365; 8 March 1907, col. 1157.
24. B. Harrison, *Prudent Revolutionaries* (Oxford, 1987) p. 223.
25. *A Suffrage Annual and Woman's Who's Who* (London: Women's Press, 1913).
26. M. Pugh, *History Students' Pamphlets* (London, 1982) p. 5.

Notes to Chapter 10: Breaking into the Professions

1. *Pharmaceutical Journal and Transactions*, 1872–3, pp. 268, 366–7, 480, 698–9, 936–41. S. W. F. Holloway, *The Royal Pharmaceutical Society of Great Britain* (London, 1991) pp. 254–64 and passim. *The Individualist*, March 1905, p. 18.
2. *Pharmaceutical Journal*, 25 January 1873, pp. 588–9. Holloway, *The Royal Pharmaceutical Society*, pp. 256–64.
3. *Journal of Royal Institute of British Architects*, 10 December 1898, p. 77; and 11 March 1899, p. 279. Lynne Walker, in J. Attfield and P. Markham (eds), *A View from the Interior: Women, Feminism and Design* (London, 1989) pp. 90ff.
4. C. M. Ford, *Aleen Cust, Veterinary Surgeon* (Bristol, 1990) passim. I. Pattison, *The British Veterinary Profession* (London, 1984) pp. 152–9.

5. *Dental History*, 23 November 1991, p. 325, article by E. M. Cohen and R. A. Cohen, 'The Autobiography of Lillian Lindsay'. Communication from Prof. Christine Hillam, University of London.
6. *The Accountant*, 9 June 1888, p. 365; 8 May 1909, pp. 587 and 645; 18 April 1914, p. 563; 26 October 1918, p. 229; 25 January 1919, p. 64; 10 May 1919, p. 398; 13 December 1919, p. 517.
7. *Solicitors' Journal*, 4 April 1914, p. 408; 14 September 1918, p. 782; 15 March 1919, p. 369; 5 April 1919, p. 403.
8. Ibid., 18 January 1919, p. 219; 5 April 1919, p. 414.
9. H. Martindale, *Servants of the State* (London, 1933) p. 16.
10. *Quarterly Review*, vol. 151 (January 1881) p. 186. *Nineteenth Century*, vol. 10 (September 1889) p. 364.
11. Hansard, *House of Commons Debates*, 5 August 1921, col. 1904. Martindale, *Servants of the State*, pp. 30, 40, 54, 61ff.
12. Martindale, *Servants of the State*, p. 64. D. Evans, *Women and the Civil Service* (London, 1934) p. 42.
13. Hansard, *House of Commons Debates*, 27 October 1919, col. 362; 5 August 1921, cols 1900 and 1913. Martindale, *Servants of the State*, p. 105.
14. S. G. Hutchinson, *History of the Royal Academy* (London, 1968) pp. 60, 138, 176, 198. D. Cherry, *Painting Women* (London, 1963) pp. 96, 236. Royal Academy Archives, Council Minutes, 11 November 1879; General Assembly Minutes, 6 January 1880; Annual Report of Council, 1879. J. G. Millais, *Life and Letters of Sir J. E. Millais* (London, 1899) vol. 1, p. 147.
15. Communication from J. C. Dagnall of the Society of Chiropodists.

Notes to Chapter 11: Observations and Reflections

1. *Hansard*, Parliamentary Debates, 20 March 1856, cols 409ff; 3 March 1857, cols 1691ff. A. Horstman, *Victorian Divorce* (London, 1985) passim.
2. L. Holcombe, *Wives and Property* (London, 1983) pp. 18, 210 and passim.
3. R. Symonds, *Oxford and Empire* (Oxford, 1991) p. 253. Gertrude Bell, *Letters*, Vol. II, London, 1927, p. 686, 717.
4. W. P. Livingstone, *Mary Slessor of Calabar* (London, 1916) and subsequent biographies.
5. R. Kipling, *The Day's Work* (London, 1927) p. 194.
6. *DNB*, J. D. Bethune.
7. A. J. Broomhall, *Hudson Taylor and China's Open Century* (1981–9) vol. IV, pp. 48, 351; vol. VII, p. 41. Peter Williams, 'The Missing Link', in F. Bowie et al. (eds), *Women and Missions Past and Present* (Oxford, 1993) pp. 43ff.
8. R. Symonds, *Oxford and Empire* (Oxford, 1991) pp. 216, 223. Bowie, *Women and Missions*, p. 66.
9. F. Engels, *Origin of the Family* (London, 1884). S. Strauss, *Traitors to the Masculine Cause* (London, 1982) p. 123.
10. R. Anderson, *Educational Opportunities in Victorian Scotland* (Oxford, 1983) p. 254.

11. F. P. Cobbe, *Life* (London, 1894) vol. II, p. 210.
12. B. King, *Life of Mazzini* (London, 1929) p. 219.
13. P. Grimshaw, *Womens Suffrage in New Zealand* (Auckland, 1972). There are biographies of all those mentioned.
14. H. W. Nevinson, *More Changes, More Chances* (London, 1925) ch. 14.
15. F. M. Leventhal, *The Last Dissenter: H. N. Brailsford and his World* (Oxford, 1985).
16. Brian Harrison, *Separate Spheres* (London, 1978).

Index